T0210435

Synthesis Lectures on Engineering, Science, and Technology

The focus of this series is general topics, and applications about, and for, engineers and scientists on a wide array of applications, methods and advances. Most titles cover subjects such as professional development, education, and study skills, as well as basic introductory undergraduate material and other topics appropriate for a broader and less technical audience.

Maik Peter Kaufmann · Bernhard Wicht

Monolithic Integration in E-Mode GaN Technology

 Springer

Maik Peter Kaufmann
Texas Instruments
Freising, Germany

Bernhard Wicht
Leibniz University Hannover
Gehrden, Germany

ISSN 2690-0300   ISSN 2690-0327 (electronic)
Synthesis Lectures on Engineering, Science, and Technology
ISBN 978-3-031-15627-4   ISBN 978-3-031-15625-0 (eBook)
https://doi.org/10.1007/978-3-031-15625-0

This Springer imprint is published by the registered company Springer Nature Switzerland AG
The registered company address is: Gewerbestrasse 11, 6330 Cham, Switzerland

Preface

The objective of this book is to provide a systematic and comprehensive insight into monolithic integration in Gallium nitride (GaN). GaN as a wide bandgap material enables compact and efficient power electronics. Since the high-voltage GaN transistor is a lateral device, it can be combined with other passive and active devices on the same die. This will enable Power Management Integrated Circuits (PMICs) that combine power with analog and digital functions. Similar to their silicon counterparts, a GaN power stage can be combined with sensing and control loop circuits. This book investigates challenges and presents solutions on system and circuit level for integrated power, analog and digital functions. An exemplary GaN technology is characterized for its analog and digital integration capabilities. Circuit design techniques are presented which exploit the GaN-specific benefits and also solve the technology-related challenges such as the lack of suitable complementary devices. This way, the book provides a fundamental resource for the relatively new field of GaN PMIC design.

With voltage capabilities in the order of 650V, GaN is well suitable for offline power conversion. The book gives an introduction into e-mode GaN-on-Silicon technologies as a new and promising technology for power converters connected directly to the AC power grid. Consequently, the book explores circuit design for power converter ICs using a 650V e-mode GaN-on-Silicon technology to achieve highly energy-efficient offline power converters with low component count and low PCB complexity in the growing field of point-of-load power converters (chargers, LED lighting, server applications, etc.). The goal is to monolithically integrate a GaN power stage together with a control loop and a HV supply regulator generating the supply voltage for the IC directly from the HV input. Therefore, the system complexity is reduced while concurrently the parasitic gate loop inductance is minimized close to zero to fully enable the superior switching characteristics of GaN transistors.

In particular, this book presents a design example of an LED driver IC in GaN operating directly from the 110V and 230V AC power grid. With 95.6% the IC shows the highest peak efficiency of converters with fully integrated power stage. Thanks to the fast-switching GaN integration the low component count and small passive component size results in a power density up to $44W/in^3$, which is the highest reported in comparison to the state of the art in this voltage and power class. The results can be transferred to other GaN converters and applications spaces.

The book deals in detail with the following main topics: (1) Introduction into monolithic GaN process technology with a focus on the similarities and differences to traditional silicon technologies. (2) The gate loop challenge with GaN transistors due to tight gate-voltage requirements and different approaches to ensure a safe operation. (3) Measurement-based characterization of technology properties for analog, digital, and mixed-signal circuit design. (4) Design and implementation of a monolithic buck converter IC in a 650V e-mode GaN technology. (5) Comparison between the designed GaN-based converter and a commercially available silicon converter based on comprehensive experimental characterization and analysis. Specifically, this includes a detailed loss analysis enabling the reader to understand the trade-offs and benefits of both technologies to implement compact and highly efficient offline power converters.

This book is based on our research at the Institute for Microelectronic Systems at Leibniz University Hannover, Hannover, Germany. We are grateful to many team members at the university as well as at our industry partners.

A special thanks goes to our families, without their love and support this book would not have been possible.

Freising, Germany   Maik Peter Kaufmann
Gehrden, Germany   Bernhard Wicht
June 2022

Contents

Acronyms

Abbreviations

2DEG	Two-Dimensional Electron Gas
AC	Alternating Current
ADC	Analog-to-Digital Converter
AlGaN	Aluminum-doped GaN
BCD	Bipolar, CMOS, and Drain extended
BCM	Boundary Conduction Mode
BSIM	Berkeley Short-channel IGFET Model
CCM	Continuous Conduction Mode
CICC	Custom Integrated Circuits Conference
CMOS	Complementary Metal–Oxide–Semiconductor
CTAT	Complementary To Absolute Temperature
DC	Direct Current
DCR	DC Resistance
DCFL	Direct Coupled FET Logic
DCM	Discontinuous Conduction Mode
d-mode	Depletion-Mode
EMI	Electromagnetic Interference
e-mode	Enhancement-Mode
ESD	Electrostatic Discharge
ESL	Equivalent Series Inductance
ESR	Equivalent Series Resistance
FET	Field-Effect Transistor
FFT	Fast Fourier Transformation
GaAs	Gallium Arsenide
GaN	Gallium Nitride
GiT	Gate Injection Transistor
HEMT	High Electron Mobility Transistor
HV	High Voltage

IC	Integrated Circuit
IEEE	Institute of Electrical and Electronics Engineers
ISSCC	International Solid-State Circuits Conference
IGBT	Isolated Gate Bipolar Transistor
LDMOS	Lateral Drain-extended Metal–Oxide–Semiconductor
LED	Light Emitting Diode
LFER	Lateral Field-Effect Rectifier
LV	Low Voltage
MIM	Metal Insulation Metal
MOSFET	Metal–Oxide–Semiconductor Field-Effect Transistor
NMOS	N-type Metal–Oxide–Semiconductor
NTC	Negative Temperature Coefficient
OCP	Over-current Protection
OTP	Over-temperature Protection
PBTI	Positive Bias Temperature Instability
PCB	Printed Circuit Board
PFC	Power Factor Correction
PMOS	P-type Metal–Oxide–Semiconductor
PTAT	Proportional to Absolute Temperature
PVT	Process-Voltage-Temperature
PWM	Pulse Width Modulation
QRM	Quasi-resonant Mode
RF	Radio Frequency
RMS	Root Mean Square
RTL	Resistor-Transistor Logic
Si	Silicon
SiC	Silicon Carbide
SMPS	Switch Mode Power Supplies
SOI	Silicon-on-Insulator
SPICE	Simulation Program with Integrated Circuit Emphasis
USB	Universal Serial Bus
UVLO	Under-voltage Lockout

Symbols

A	Area (m^2)
A_{int}	Intrinsic Gain (1)
A_V	Voltage Gain (1)
BFOM	Baliga's Figure-of-Merit (V)

BVDSS   Drain-Source Breakdown Voltage (V)
C_{aux} Auxiliary Capacitance (F)
C_{bias} Bias Capacitor (F)
C_{bp} Bypass Capacitor (F)
C_{bulk} Bulk Input Buffer Capacitance (F)
C_{Dfw} Capacitance of the Freewheeling Diode (F)
C_{ds} Drain-Source Capacitance (F)
C_{EMI} EMI Filter Capacitance (F)
C_{gd} Gate-Drain Capacitance (F)
C_{gg} Total Gate Capacitance (F)
c_{gs} Small Signal Gate-Source Capacitance (F)
C_{gs} Gate-Source Capacitance (F)
C_{hv} High-Voltage Capacitance (F)
C_{Lout} Capacitance of the Output Inductor (F)
$C_{O(ER)}$ Energy-Related Output Capacitance (F)
C_{oss} Transistor Output Capacitance (F)
C_{out} Output Capacitance (F)
C_{par} Parasitic Capacitance (F)
C_{pcb} Parasitic Capacitance of PCB Traces (F)
C_{plate} Capacitance of a Plate Capacitor (F)
C_{probe} Probe Capacitance (F)
C_{sw} Capacitance at the Switching Node (F)
D Damping Factor (1)
d Distance (m)
E Electric Field Strength (V m^{-1})
E_{Cr} Critical Electric Field (MV cm^{-1})
E_{Csw} Energy of the Switching Node Capacitance (J)
E_G Bandgap (eV)
ϵ_0 Absolute Permittivity (F m^{-1})
ϵ_r Relative Permittivity (1)
f Frequency (Hz)
f_{line} Frequency of the AC Line (Hz)
f_{osc} Oscillator Frequency (Hz)
f_{sw} Switching Frequency (Hz)
g_m Small Signal Transconductance (S)
I_{bias} Bias Current (A)
I_D Drain Current (A)
I_{DD} Supply Current (A)
$I_{DD,off}$ Supply Current during the OFF-time (A)
$I_{DD,on}$ Supply Current during the ON-time (A)
I_{dg} Drain Gate Current (A)

I_{in}	Input Current (A)
I_L	Inductor Current (A)
$I_{L,peak}$	Maximum Inductor Current Value (A)
I_{out}	Output Current (A)
I_{peak}	Maximum Current Value (A)
I_{valley}	Minimum Current Value (A)
i_{sns}	Sensed Current (A)
I_{sat}	Saturation Current (A)
L	Transistor Length (m)
L_{EMI}	EMI Filter Inductance (H)
L_{out}	Output Inductance (H)
L_p	Parasitic Inductance (H)
μ_n	Electron Mobility (cm^{2}V^{-1} s^{-1})
P_{in}	Input Power (W)
P_{out}	Output Power (W)
P_{sw}	Switching Losses (F)
Q_g	Gate Charge (C)
Q_{gs}	Gate-Source Charge (C)
Q_{gd}	Gate-Drain Charge (C)
Q_{oss}	Output Charge (C)
Q_{rr}	Reverse Recovery Charge (C)
Q_{sw}	Switching Charge (C)
R_{bias}	Bias Resistor (Ω)
R_{drv}	Driver Resistance (Ω)
r_{Dfw}	Differential Resistance of the Freewheeling Diode (Ω)
r_{ds}	Small Signal Drain-Source Resistance (Ω)
R_{DS}	Drain-Source Resistance (Ω)
$R_{DS,on}$	Drain-Source on-Resistance (Ω)
R_f	Fusible Resistor (Ω)
R_G	Gate Resistor (Ω)
r_{gs}	Small Signal Gate-Source Resistance (Ω)
r_{gd}	Small Signal Gate-Drain Resistance (Ω)
r_{load}	Small Signal Load Resistance (Ω)
R_{Lout}	DC Resistance of the Output Inductor (Ω)
R_{PU}	Pull-Up Resistor (Ω)
R_S	Source Resistance (Ω)
R_{shunt}	Shunt Resistance (Ω)
R_{sp}	Specific Resistance (Ω mm^2)
S_{ID}	Drain Current Spectral Noise Density (A^2 Hz^{-1})
T	Cycle Time (s)
T_{abs}	Absolute Temperature (K)

$t_{d,pc}$	Delay Time of the Peak Current Detection (s)
$t_{d,zc}$	Delay Time of the Zero-Current Detection (s)
t_f	Fall Time (s)
t_I	Time with $I > 0$ (s)
t_{off}	Off-Time (s)
t_{on}	On-Time (s)
t_{pc}	Moment of Peak Current Detection (s)
t_{PD}	Propagation Delay (s)
$t_{PD,off}$	Propagation Delay for Turn-Off (s)
$t_{PD,on}$	Propagation Delay for Turn-On (s)
T_{period}	Period Time (s)
t_{zc}	Moment of Zero Current (s)
V	Voltage (V)
V_{bias}	Bias Voltage (V)
V_{BS}	Backgate-Source Voltage (V)
V_{DD}	Supply Voltage (V)
V_{DDL}	Low-Side Supply Voltage (V)
V_{Dfw}	Forward Voltage of the Freewheeling Diode (V)
V_{DS}	Drain-Source Voltage (V)
V_f	Diode Forward Voltage (V)
V_G	Gate Voltage (V)
V_{GD}	Gate-Drain Voltage (V)
V_{GS}	Gate-Source Voltage (V)
V_{in}	Input Voltage (V)
v_{in}	Small Signal Input Voltage (V)
V_{line}	AC Line Voltage (V)
V_{offset}	Offset Voltage (V)
V_{osc}	Oscillator Voltage (V)
V_{out}	Output Voltage (V)
V_{ovd}	Overdrive Voltage (V)
v_{out}	Small Signal Output Voltage (V)
V_{ref}	Reference Voltage (V)
V_S	Source Voltage (V)
v_{sat}	Saturation Velocity (cm s^{-1})
v_{sns}	Sensed Voltage (V)
V_{sw}	Voltage at the Switching Node (V)
V_{th}	Threshold Voltage (V)
W	Transistor Width (m)

Introduction

1

Offline power converters are an integral part of various electronic devices in today's world. They are utilized as chargers for portable devices like smart phones and laptops as well as for direct power supplies in office and home electronics such as television, desktop computers, screens, and energy-efficient LED lighting. Furthermore, power converters are necessary in motor drive applications for washing machines, industrial robots, and e-mobility. Servers and server farms as backbone of the connected world require power electronics to convert the voltages above 100 V from the power grid to 1 V and less for the processors, memory, and hard drives integrated in servers. An exemplary overview of the various fields of application for offline power converters is depicted in Fig. 1.1. For all of these applications, high conversion efficiency is desired in order to reduce the cost for electricity and also to reduce the global power consumption as a part of the effort to slow down and stop climate change. In most applications, increasing the conversion efficiency has a direct influence on the power consumption. However, for high-density electronics such as server farms high power conversion efficiency is even more important since it reduces the cooling effort and thereby adds a secondary effect by lowering the energy consumption for the cooling system.

Besides high efficiency, also high power density is desired in all applications in order to reduce the volume of the power converters. Thereby, for instance, chargers of personal electronics become smaller and lighter. In the example of e-mobility, smaller and lighter power electronics save valuable space and reduce the total weight, which in turn can be utilized to increase the energy storage capacity for an extended range. The power density of converters is mostly limited by the size of passive energy storage components, namely, buffer capacitors and power inductors or transformers. The size of some capacitors and especially of inductive components can be significantly reduced by increasing the operation frequency of power converters. For input voltages of 400 V and higher, traditional silicon-based power switches such as the Isolated Gate Bipolar Transistor (IGBT) and the superjunction MOSFETs reach their limits at switching frequencies of around some 20 and 80 kHz, respectively.

© The Author(s), under exclusive license to Springer Nature Switzerland AG 2022
M. P. Kaufmann and B. Wicht, *Monolithic Integration in E-Mode GaN Technology*,
Synthesis Lectures on Engineering, Science, and Technology.
https://doi.org/10.1007/978-3-031-15625-0_1

Fig. 1.1 Exemplary application fields for offline power converters, credits from top-left to bottom-right: zhu difeng/Shutterstock.com, Oleksandr_ Delyk/Shutterstock.com, Phonlamai Photo/Shutterstock.com, Nerthuz/Shutterstock.com, Oleksiy Mark/Shutterstock.com

Due to superior switching performance, power transistors based on wide bandgap materials such as gallium nitride (GaN) and silicon carbide (SiC) enable higher switching frequencies while maintaining good power efficiency. Thereby, new operating schemes with smaller passive components are possible, which cannot be achieved by utilizing silicon power transistors. Of the established wide bandgap materials, GaN shows the best Baliga's figure-of-merit (BFOM), which indicates the smallest transistor size for a defined breakdown voltage and on-state resistance [1]. Furthermore, the smaller size leads to smaller specific capacitances. Together with the absence of reverse-recovery native to GaN transistors, this leads to efficient high-frequency switching operation in the MHz range. In addition, the lateral structure of GaN power transistors allows for monolithic integration of power converter Integrated Circuit (IC)s and thus enables the design of compact and easy-to-use power modules.

With the ongoing development of GaN-on-silicon (Si) technology, the cost of GaN power transistors reduces and they are commercially available at prices starting below 1 $ [2]. Due to the superior performance and the expected cost reduction when the global production capacity for GaN transistors increases, market analysts recently predicted a year-to-year annual growth rate above 75% for the next several years leading to a global market volume of more than 700 million dollars by the year 2025 [3]. This is backed by various press announcements on shipped GaN devices by companies active in the GaN power semiconductor industry. In September 2019, Power Integrations had one million GaN-based converter ICs [4] shipped to Anker for a high power USB charger [5]. In January 2021, Navitas announced 13 million units shipped [6], followed by 20 million units sold by GaNSystems in February 2021 [7]. One indicator for a prosperous future of GaN is the fact that GaN power ICs entered the high-volume USB-C charger market leading to chargers with unprecedented power density of 16 W/in^3 combined with high efficiency of 92.5% for a 100 W dual USB-C PD charger [8]. Since mid-2020s, Navitas announced various cooperations with well-established players in the consumer and personal electronics market such as Lenovo [9], Oppo [10], Dell [11], Xiaomi [12], LG [13], and Samsung [14]. The broad adoption of GaN for USB-C chargers as a mass market is generally viewed as a milestone leading to further cost reduction of GaN. Thereby, also other markets such as server and data center supplies [15] are enabled. Furthermore, with the increasing markets for GaN, more investments into research and development of the GaN process technology can be expected. In addition, the integration capability of GaN may lead to higher levels of integration [16], thus paving the path towards compact and easy-to-use power converters with high efficiency.

Figure 1.2 shows the integration trend for offline power converters in the 10 to 50 W power range using different topologies. It depicts peak efficiency values of various converter implementations between the years 2010 and 2020, which have been presented at IEEE conferences or in datasheets or reference design reports. The graph includes some power converter ICs integrated monolithically on one silicon die [17–19] (filled circles) achieving efficiencies between 89 and 91%. The converters employing discrete GaN transistors as power switch [20–22] (marked with triangles) achieve higher power efficiency than a silicon implementation published in the same year. Even higher efficiency above 94% is achieved by GaN ICs integrating the high-voltage power transistor as well as the gate driver on one GaN die [23, 24] (marked with squares). These two converters use the active clamp flyback topology and support the highest power level of the converters shown in this figure of up to 45 W and 65 W, respectively. The highest peak efficiency of 95.6% is reported for a monolithically integrated GaN converter IC (marked with a diamond), which also includes an analog control loop as well as a supply regulator [25]. Concurrently, a higher level of integration in one IC reduces the system complexity and additionally improves the reliability of the solution due to lower component count. Thereby, also lower cost can be achieved on system level.

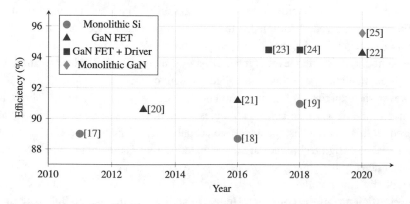

Fig. 1.2 Development of peak efficiency and integration level from silicon to GaN for 10 to 50 W converters supplied by 110 to 230 AC line voltage

Besides the financial aspect, increasing the power supply efficiency is an important contribution to reduce the global energy consumption and associated greenhouse gas emissions. The energy consumption of data centers alone is estimated to be about 200 TWh for the year 2018, which is 1% of the global electric energy consumption [26]. Even more significant, electrical lighting accounts for 15% of the global electric energy consumption and 5% of global greenhouse gas emissions [27]. While these values are likely to reduce with further adoption of LED lighting, increasing the efficiency of the required power converter has an impact on cutting global energy consumption and greenhouse gas emission. Thus, this work provides an investigation into monolithic integration in GaN technology in order to explore the chances and limitations for compact and efficient offline power supplies in the future.

1.1   Scope of This Book

The scope of this work is summarized and depicted in Fig. 1.3. It is based on the trend towards smaller power supplies with higher efficiency by using GaN technology. Thus, it contributes to the development of solutions for reducing the global electric power consumption as well as the overall cost of offline power converters.

In this work, a power supply converter for LED lighting is utilized as investigation vehicle illustrated in Fig. 1.3a. Most considerations and results of this work are also valid for other offline power converters such as USB chargers and wall adapters (Fig. 1.3b). In general, these applications require operation of the converter at both, 110 and 230 V, power grids. High efficiency and power density are desired, leading to compact power supplies and contributing to reduce the global power consumption with its associated greenhouse gas emissions. Furthermore, a high level of integration is preferred to obtain an easy-to-use system with limited printed circuit board (PCB) complexity as well as to reduce the required number of components in high-volume markets such as LED lighting.

In order to address all of these requirements, this work aims for a monolithic power converter IC in GaN technology. With the utilization of GaN, the required input voltage capabilities are achieved. Additionally, the low specific capacitances of GaN enable higher switching frequencies and, consequently, higher power density due to lower passive energy storage components. A high level of integration in GaN is expected to lead to an easy-to-use system implementation with low external component count.

However, the experiences for integration in GaN are very limited and the fabrication process is not as mature as for silicon technologies, yet. That poses some of the main challenges addressed by this work. Thus, the integration capabilities of the utilized GaN technology as well as suitable design techniques are investigated as two major topics. In order to achieve high power density, high operation frequencies are required of the power converter. In combination with high input voltages, this generally leads to significant switching losses impairing the power efficiency of the converter. Hence, in addition to the use of GaN technology for the power switch, careful system-level considerations are required to maximize the achieved

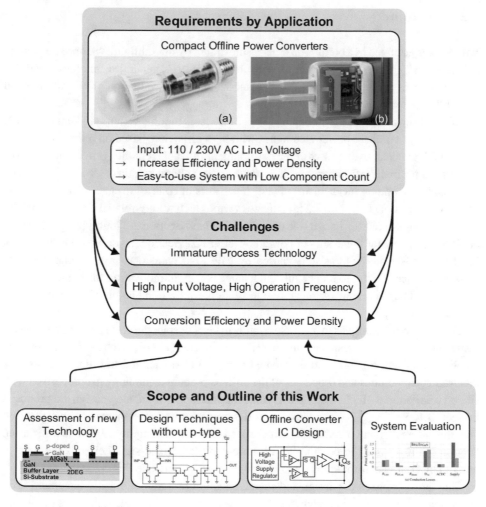

Fig. 1.3 Summary of this work, **a** courtesy of Fraunhofer IAF [28], **b** courtesy of Navitas Semiconductor [29]

conversion efficiency. Furthermore, the circuits utilized for the converter IC need to be optimized for low power consumption. An extensive system evaluation is performed to allow the considerations and results of this work to be transferred to other applications. A comparison to a silicon-based LED power supply converter is performed in order to quantify the benefits achieved by an integrated GaN solution.

1.2   Outline

This book is arranged in a bottom-up order as illustrated by the subfigure "Scope and Outline of this Work" in Fig. 1.3 from left to right. It starts with a general technology overview and assessment leading to the design, characterization, and evaluation of an experimental monolithic buck converter IC.

After the introduction in this chapter, Chap. 2 discusses the chances and challenges of GaN technology for compact power electronics with high efficiency based on a literature study. Section 2.1 gives an overview over the development in GaN technology, in particular, how the desired normally-off enhancement-mode (e-mode) characteristic can be achieved. The fundamental operation mechanisms of GaN transistors in contrast to state-of-the-art silicon devices are discussed. Section 2.2 investigates various technology-related challenges, which have to be considered for integration of circuits in GaN. Section 2.3 highlights the benefits of GaN especially for the design of power converter ICs. Section 2.4 derives the challenges for driving GaN power transistors caused by gate loop parasitics, in particular, the gate loop inductance. Trying to reduce this inductance is a main motivation for higher levels of integration. Thus, a review of integration levels achieved around and in GaN technologies is given in Sect. 2.5.

Concepts and techniques to integrate circuits in GaN despite the various challenges posed by the technology are investigated in Chap. 3. Fundamental circuits such as common-source amplifiers, current mirrors, and differential pairs are characterized in order to assess the integration capabilities and properties of the utilized process technology in comparison with silicon technologies. Techniques to overcome the challenges for analog design are proposed in Sect. 3.1. Section 3.2 looks into digital design without complementary transistors and how this affects critical parameters such as speed, power consumption, and area utilization. Section 3.3 discusses the possibilities of more complex mixed-signal circuits required to implement analog control loops for power converters to conclude this chapter.

Chapter 4 discusses the design and characterization of a monolithically integrated buck converter IC on circuit and system level. A power converter for LED lighting is employed as an exemplary investigation vehicle in order to assess the possibilities for monolithic integration in GaN. Thus, general system-level considerations for this application such as input and output specifications as well as operation mode and control method are discussed in Sect. 4.1. Section 4.2 presents the first integration step with high-voltage power transistor and the required gate driver including characterization results for both of them. Section 4.3 introduces the implementation of the analog control loop according to the considerations in the beginning of Chap. 4. Design techniques to enhance the performance of required circuits are presented and characterized. In order to support self-biased offline operation, a high-voltage supply regulator is included on the IC. Its operation principle is presented in Sect. 4.4 alongside a characterization based on transient waveforms. The chapter is concluded with a top-level characterization of the IC implemented as a low-side buck converter with constant current output in Sect. 4.5. The performance in terms of efficiency over input voltage as well as the interaction of all subcircuits is provided.

Chapter 5 presents a detailed characterization along with a comparison between the implemented GaN IC and a commercially available silicon converter IC with similar input and output ratings. The silicon-based IC is introduced in Sect. 5.1 with measured top-level waveforms and a comparison of the semiconductor chips. The switching loss is one main power loss mechanism of both ICs. Thus, a model of the various capacitances at the switching node is introduced and validated by transient measurements in Sect. 5.2. Section 5.3 investigates the power loss distribution based on the different loss mechanisms as well as the components required to implement a buck converter. Section 5.4 provides an efficiency comparison for various input voltages and operation frequencies as well as different power densities.

Chapter 6 summarizes and concludes the results of this work.

1.3   Highlights of This Book

This section summarizes how this work contributes to the goal of compact and easy-to-use offline power converters with high efficiency by integration in e-mode GaN-on-Si technology as outlined in Fig. 1.3.

Monolithic GaN power converter IC. This work presents the first GaN-integrated power converter IC with an analog control loop suitable for operation directly for the AC mains. It provides a constant current output at power levels up to 29 W for a wide input voltage range between 85 to 400 V. Hence, it is suitable for both, 110 and 230 V, power grid. The design achieves a peak efficiency of 95.6% and maintains high efficiency above 90% for input voltages above 300 V. Compared with state-of-the-art silicon solutions, the power losses are reduced by up to 42% due to circuit design optimized for low power consumption as well as the efficient switching of GaN. Additionally, higher power density is achieved by exploiting the faster switching capabilities of GaN with a suitable operation mode as well as by a high level of integration reducing the numbers of external components required on PCB. The converter IC was first published at ISSCC 2020 [25] and, subsequently, invited for a detailed article in the December 2020 issue of the IEEE Journal of Solid-State Circuits [30].

Analog and mixed-signal circuit design in GaN technology. This work provides an in-detail assessment of today's GaN-on-Si technology for analog, digital, and mixed-signal integration. It compares key figures-of-merit such as intrinsic and single-stage gain of analog amplifier stages (27 dB for GaN vs. 45 dB for silicon) as well as power-delay-product of digital logic gates (4.2×10^{-13} Ws for GaN vs. 4.5×10^{-17} Ws for silicon). Fundamental characteristics such as transistor matching and noise are investigated and all results are supported by measurements. Where applicable, the characteristics are related to known challenges caused by the GaN fabrication process technology. Design techniques to compensate these effects on circuit and system level such as bootstrapping and auto-zeroing are introduced and evaluated. With the proposed auto-zero loop, the mean input referred offset of a peak current comparator can be reduced from initially 70 mV (standard deviation) to less than 5 mV. In this work, a new level of integration in e-mode GaN technology is achieved. While power stages and digital circuits in GaN have been demonstrated before, various analog and mixed-signal circuits presented in the following chapters represent pioneering work in the field of monolithic integration in GaN. On the path towards compact and efficient offline power converters, this work demonstrates the first analog control loop. Key circuits, such as a single-supply rail-to-rail gate driver with a propagation delay of 80 ns as well as an auto-zeroed comparator for cycle-by-cycle peak current control, are central components for power converter ICs in GaN.

High-voltage functions in GaN technology. Besides the high-voltage power transistor, the design in this work integrates diverse other high-voltage functions. Connected to the power transistor, an on-chip 100 fF linear, high-voltage capacitor suitable for 400 V operation is proposed in this work for efficient close-proximity sensing in the power stage. It is employed as part of the integrated analog control loop to ensure operation in quasi-resonant mode (QRM).

In order to support self-biased offline operation without increasing the system complexity by additional components such as an auxiliary winding or an external supply regulator, a high-voltage supply regulator is proposed, implemented, and characterized. It generates the supply voltage of typical 6 V with less than 50 mV ripple for all circuits integrated on the GaN die directly from the high-voltage switching node and is itself part of the IC. To charge an auxiliary capacitor buffering the supply voltage, a current up to 8.5 mA can be provided temporarily. This is sufficient to support the required supply current of max. 830 μA for the full converter IC with a duty cycle of less than 20% for the supply regulator.

Loss analysis and comparison to a silicon converter IC. With smaller die area (2.1 mm^2 vs. 5.8 mm^2) and smaller specific capacitances, the GaN implementation enables faster switching operation and lower on-resistance $R_{DS,on}$ while maintaining higher efficiency than a state-of-the-art silicon converter. In a detailed loss analysis and comparison, the core losses of the power inductor as well as various conduction losses are identified as main contributions to the total power loss of the converters. At low input voltages, the lower ($R_{DS,on}$) of the GaN power transistor significantly contributes to achieve 2.3% higher efficiency than the silicon implementation. At high input voltages, the GaN IC reaches 2.6% higher efficiency mainly due to the efficient circuits designed in the GaN technology leading to lower power consumption for the proposed IC. Additionally, the faster switching capabilities of GaN enable the use of a broader selection of inductors. The dominant core losses of the converter can be reduced by employing inductors not supported by the silicon IC. On top of that, 7.8% higher power density can be achieved by selecting inductors in a smaller package, which are not supported by the silicon implementation, either.

Operating modes and frequencies of GaN power converters. The use of GaN transistors for power converters is often associated with operation frequencies of 1 MHz and higher (e.g. [31]). While such switching frequencies and associated high power densities can be achieved with GaN, the viability with respect to the conversion efficiency heavily depends on the specific application and its input and output conditions. In this work, various operation modes and switching frequencies are investigated. It is demonstrated that for 10 to 30 W, 50 V non-isolated output, operation in QRM is most beneficial to achieve high efficiency over the full input voltage range 85 to 400 V for offline power converters. The characterization across various switching frequencies and input voltages proves that the best efficiency is achieved at switching frequencies up to 150 kHz. Higher switching frequencies are possible in order to achieve higher power density. However, frequencies above 300 kHz come at the cost of significantly impaired efficiency with limited gain in power density especially at input voltages above 300 V. Thus, operation in QRM with switching frequencies below 300 kHz is recommended for the investigated application space with power levels below 30 W at 50 V output voltage.

Gate loop challenge for GaN drivers. The combination of high slew rates and tight gate voltage requirements for e-mode GaN power transistors poses considerable challenges to the required gate driver. Applicable and comprehensive equations are derived in order to estimate maximum allowable parasitics in the gate loop. In order to avoid a destructive gate voltage overshoot when the power transistor is turned on by an unipolar gate driver, a series gate resistance greater than 5 Ω is required for an exemplary GaN transistor. At the same time, the maximum allowable gate resistance for this transistor is 3.8 Ω to prevent unwanted turn-on of the transistor during slewing of its drain-source voltage. Furthermore, for the same purpose the maximum tolerable gate loop inductance is as low as 1 nH. Thus, discrete gate drivers for GaN transistors require additional protection such as clamps and additional gate-source capacitors or have to be implemented as bipolar or multi-level gate drivers providing negative gate-source voltages for turn-off. Avoiding this complexity and associated cost is one major motivation for integrating the driver and the power transistor on one GaN die. These considerations have been published at CICC 2020 [32].

References

1. Lidow, A. et al. (2012). *GaN transistors for efficient power conversion* (1st ed., p. 208). El Segundo, CA: Power Conversion Publications. ISBN: 9780615569253.
2. GaN Systems. (2021). *GaN Power Transistor Prices Drop Below* $ 1.00. Retrieved April 14, 2021, from https://gansystems.com/newsroom/gan-under-one-dollar/.
3. Developpement, Y. (2020). *Power GaN device market forecast between 2019 and 2025*. Retrieved April 14, 2021, from http://www.yole.fr/iso_upload/News_Illustration/2020/ILLUS_CS_MONITOR_Q4-2020_Power_GaN_DeviceMarketForecast_YOLE_Dec2020.jpg.
4. Businesswire. (2019). *Power Integrations Delivers One-Millionth GaN-Based InnoSwitch3 IC*. Retrieved April 14, 2021, from https://www.businesswire.com/news/home/20190929005027/en/.
5. SystemPlus Consulting. (2019). *GaN-on-Sapphire HEMT Power IC by Power Integrations— InnoSwitch3 Flyback Switcher Power IC in Anker PowerPort A2017 (Sample)*. Retrieved

April 06, 2021, from https://www.systemplus.fr/wp-content/uploads/2019/07/SP19480-Power-Integrations-GaN-on-Sapphire-HEMT_sample.pdf.

6. Navitas. (2021). *Navitas Ships 13,000,000 GaNFast Power ICs with World-Class Reliability.* Retrieved April 14, 2021, from https://www.navitassemi.com/navitas-ships-13000000-ganfast-power-ics-with-world-class-reliability.

7. GaN Systems. (2021). *GaN Systems Ships 20,000,000 GaN Transistors.* Retrieved April 14, 2021, from https://gansystems.com/newsroom/20-million-transistorsshipped/.

8. GaN Systems. (2021). *The Industry's Smallest and Smartest 100 W GaN Charger Reference Design.* Retrieved April 14, 2021, from https://gansystems.com/newsroom/smallestsmartest-100w-gan-charger/.

9. Navitas. (2020). *Lenovo Partners with Navitas Again to Deliver theWorld's First GaN-Fast 90 W Fast Charger for E-sports Mobile Phones.* Retrieved April 14, 2021, from https://www.navitassemi.com/lenovo-partners-with-navitas-again-todeliver-the-worlds-first-ganfast-90w-fast-charger-for-esports-mobile-phones/.

10. Navitas. (2020). *Navitas GaN IC Drives OPPO's New Generation of Fast Charging.* Retrieved 14, 2021, from https://www.navitassemi.com/navitas-gan-ic-drivesoppos-new-generation-of-fast-charging/.

11. Navitas. (2020). *Dell Adopts Navitas GaNFast Technology for Laptop Fast Charger.* Retrieved 14, 2021, from https://www.navitassemi.com/dell-adopts-navitasganfast-technology-for-laptop-fast-charger/.

12. Navitas. (2021). *Navitas Goes Global in Xiaomi's Mi 11 Fast Charger.* Retrieved April 14, 2021, from https://www.navitassemi.com/navitas-goes-global-in-xiaomismi-11-fast-charger/.

13. Navitas. (2021). *LG Electronics Adopts Navitas GaNFast™ for World's Lightest Laptop.* Retrieved 14, 2021, from https://www.navitassemi.com/lg-electronicsadopts-navitas-ganfast-for-worlds-lightest-laptop/.

14. Navitas. (2021). *Navitas Drives Spigen's ArcStation Pro 45W: World's Smallest Samsung S21 Ultra Fast Charger.* Retrieved April 14, 2021, from https://www.navitassemi.com/navitas-drives-spigens-arcstation-pro-45w-worlds-smallest-samsung-s21-ultra-fast-charger/

15. Lidow, A. (2021). *GaN for High Density Servers.* Retrieved 14, 2021, from https://library.myebook.com/electronicspecifier/power-issue-1-february-2021/3143/%5C#page/26

16. Lidow. (2021). *How GaN Integrated Circuits Are Redefining Power Conversion.* Retrieved 14, 2021, from https://www.powerelectronicsnews.com/how-gan-integrated-circuits-are-redefining-power-conversion/.

17. Hwang, J. T. et al. (2011). A simple LED lamp driver IC with intelligent Power-Factor correction. In *2011 IEEE International Solid-State Circuits Conference* (pp. 236–238). https://doi.org/10.1109/ISSCC.2011.5746299.

18. Power Integrations. (2016). Single-Stage LED driver IC with combined PFC and constant current output for buck topology. In: LYT1402-1604 LYTSwitch-1 *Family Datasheet.*

19. Power Integrations. (2018). Reference design report for a 40 W power supply using InnoSwitch 3-Pro INN3377C-H301 and Microchip's PIC16F18325 Microcontroller. In *InnoSwitch3-Pro Reference Design RDR-641.*

20. Bandyopadhyay, S. et al. (2013). 90.6% efficient 11 MHz 22 W LED driver using GaN FETs and Burst-Mode controller with 0.96 power factor. In *2013 IEEE International Solid-State Circuits Conference Digest of Technical Papers* (pp. 368–369). https://doi.org/10.1109/ISSCC.2013.6487773.

21. Faraci, E. et al. (2016). High efficiency and power density GaN-Based LED driver. In *2016 IEEE Applied Power Electronics Conference and Exposition (APEC)* (pp. 838–842).

22. Power Integrations. (2020). 60 W power supply using InnoSwitch™3-CP PowiGaN™ INN3270C-H203. In: *DER-917 Design Example Report.*

23. Xue, L., & Zhang, J. (2017). Active clamp Flyback using GaN power IC for power adapter applications. In: *2017 IEEE Applied Power Electronics Conference and Exposition (APEC)* (pp. 2441–2448.)

24. Xue, L., & Zhang, J. (2018). Design considerations of Highly-Efficient active clamp flyback converter using GaN power ICs. In: *2018 IEEE Applied Power Electronics Conference and Exposition (APEC)* (pp. 777–782).

25. Kaufmann, M. et al. (2020). 18.2 A monolithic E-Mode GaN 15 W 400 V offline Self-Supplied hysteretic buck converter with 95.6% efficiency. In *2020 IEEE International Solid-State Circuits Conference—(ISSCC)*, San Francisco, CA (pp. 288–290).

26. Masanet, E. et al. (2020). Recalibrating global data center Energy-use estimates. In: *Science, 367.6481*, 984–986. ISSN: 0036-8075. https://doi.org/10.1126/science.aba3758. https://science.sciencemag.org/content/367/6481/984.full.pdf. https://science.sciencemag.org/content/367/6481/984.

27. de Boer, J. (2019). *Daylighting of Non-Residential Buildings Position Paper*. Retrieved April 15, 2021, from https://task50.iea-shc.org/Data/Sites/1/publications/IEASHC-Daylighting-Non-Residential-Buildings-Position-Paper.pdf.

28. LED lamps: Less energy, more light. (2014). Retrieved April 22, 2021, from https://www.fraunhofer.de/en/press/research-news/2014/march/led-lamps.html.

29. Navitas. (2021). *Navitas Drives Spigen's ArcStation Pro 45 W: World's Smallest Samsung S21 Ultra Fast Charger*. Retrieved April 22, 2021, from https://www.navitassemi.com/navitas-drives-spigens-arcstation-pro-45w-worlds-smallest-samsung-s21-ultra-fast-charger/.

30. Kaufmann, M., & Wicht, B. (2020). A monolithic GaN-IC with integrated control loop achieving 95.6% efficiency for 400 V offline buck operation. *IEEE Journal of Solid-State Circuits, 55.12*, 3446–3454.

31. Sarrafin-Ardebili, F., Allard, B., & Crebier, J. C. (2015). Analysis of Gate-Driver Circuit requirements for H-Bridge based converters with GaN HFETs. In: *2015 17th European Conference on Power Electronics and Applications (EPE'15 ECCE-Europe)* (pp. 1–10). https://doi.org/10.1109/EPE.2015.7309462.

32. Kaufmann, M., Seidel, A., & Wicht, B. (2020). Long, short, Monolithic—The gate loop challenge for GaN drivers: Invited paper. In *2020 IEEE Custom Integrated Circuits Conference (CICC)*, Boston, MA (pp. 1–5).

Fundamentals on GaN Technology for Integration of Power Electronics

2

Due to superior performance, GaN transistors experience a growing interest in the area of power electronics. From an application point of view, the first-order model and behavior of GaN power transistors are similar to silicon power transistors. The GaN transistor can be viewed as a three-terminal device, which are named gate, source, and drain. A control voltage V_{GS} between two of the terminals, gate and source, modulates the resistance R_{DS} between the last terminal, drain, and the source. If the control voltage V_{GS} turns on the transistor, a current flow can be supported in both directions: from drain to source as well as from source to drain. For power switching applications, the dynamic range of R_{RDS} is typically larger than seven decades between on- and off-state. Also, the figures-of-merit to compare different transistors are the same for GaN and silicon:

- The specific on-resistance, i.e. on-resistance normalized to the transistor area $R_{DS,On} \cdot$ area (Ωmm^2).
- The turn-on gate charge normalized to the on-resistance $Q_g \cdot R_{DS,On}$ (ΩnC).
- The transistor's output charge normalized to the on-resistance $Q_{oss} \cdot R_{DS,On}$ (ΩnC).

However, the fundamental mechanisms enabling the operation as transistor are different for GaN and silicon. To some extent, this affects the application of GaN transistors. To a much larger extent, the different operation mechanisms have a strong influence on the circuit design in GaN technology. Therefore, insights into the GaN technology as well as the relevant characteristics and mechanisms for circuit design in GaN are provided in this chapter.

Section 2.1 covers the historical development of GaN transistors as well as their key parameters for the application in switched mode power converters. The technology basics of challenges and benefits for both the design in GaN and the application of GaN in power converters are discussed in Sects. 2.2 and 2.3. With faster switching and higher slew rates of GaN transistors, parasitics in the gate loop have stronger influence. This is explored in

© The Author(s), under exclusive license to Springer Nature Switzerland AG 2022
M. P. Kaufmann and B. Wicht, *Monolithic Integration in E-Mode GaN Technology*,
Synthesis Lectures on Engineering, Science, and Technology,
https://doi.org/10.1007/978-3-031-15625-0_2

Sect. 2.4 where the challenging requirements for gate drivers applicable for GaN transistors are examined. This chapter is concluded with an investigation of the reasons for and benefits of integrating functionality in GaN. Additionally, a review of published integration levels in GaN technology is provided in Sect. 2.5. The implications of the technology-related challenges and benefits on circuit and system design techniques are discussed in Chaps. 3 and 4.

2.1   Operation Principle of GaN Transistors

The first high electron mobility transistor (HEMT) based on a GaN/aluminum-doped GaN (AlGaN) heterojunction has been reported in 1993 [1]. It shows promising characteristics such as large bandgap, high electron mobility, and high cutoff frequency due to low specific capacitances. However, it is a depletion-mode (d-mode) device, which requires a negative gate-source voltage V_{GS} to actively turn it off. Three years later, the same research group presented the first e-mode GaN field-effect transistor (FET) [2], which requires a slightly positive V_{GS} of 50 mV to be turned on.

Due to their good frequency behavior, GaN transistors were quickly adopted for radio frequency (RF) amplifiers [3–5] and are still a prominent choice for applications like cellular and wireless base stations [6–8].

Growing GaN as single crystal is very challenging [9]. Thus, the fabrication of transistors on a native substrate is not yet an affordable option. Hence, today's commercially available GaN transistors are typically fabricated on foreign substrates. First, sapphire substrates are used as mechanical carriers [1, 2], which seems to be also an economically viable option until today [10]. Later, also SiC substrates are utilized [11]. Both options, sapphire and SiC, suffer from comparably high cost and small wafer sizes. Ongoing developments show GaN grown on silicon wafers (GaN-on-Si) [12], which are widely used and much more affordable than sapphire or SiC wafers. Consequently, the interest of using GaN in mass markets for power electronics has grown. Technology research for GaN extended to improve the behavior and reliability of e-mode GaN HEMTs for high-voltage (HV) switching applications [13–16]. A detailed introduction into the GaN process technology and the fabrication of AlGaN/GaN transistors is provided in [17] and [18].

This work focuses on monolithic integration in GaN technology for power converters operated directly at the AC grid. Thus, the general technology characteristics relevant for high-voltage switching operation of GaN HEMTs are presented in this section. In general, normally-off behavior is desired for transistors employed as power switches. Different methods to achieve this behavior for GaN transistors are presented. The desired low specific on-resistance $R_{DS,On}$ of GaN transistors is mainly achieved due to the presence of a two-dimensional electron gas (2DEG) with high carrier density and electron mobility at the AlGaN/GaN interface. It forms the conductive channel for both transistors and resistors in GaN and, hence, is the origin of the device characteristics. The last part of this section provides a brief presentation of the 2DEG and the temperature behavior of its resistance and current-carrying capabilities.

Properties of GaN Transistors for Power Electronics

GaN transistors show several characteristics desirable for their application in high-voltage power converters. In comparison with Si and SiC technologies, GaN achieves superior switching figures-of-merit $R_{DS,On} \cdot Q_g$ and $R_{DS,On} \cdot Q_{sw}$. This is due to the intrinsic device properties of GaN HEMTs. The comparison of these properties is summarized in Table 2.1 for exemplary silicon, SiC, and GaN technologies.

GaN possesses a higher bandgap E_G and shows a larger critical electric field (E_{Cr}) than silicon or SiC. Therefore, the same drain-source breakdown voltage (BVDSS) can be achieved with a much shorter device length. Additionally, GaN shows a higher electron mobility (μ_n) and saturation velocity (v_{sat}), which improves the specific resistance of GaN HEMTs. Based on the investigations in [23], the relation between the achievable specific $R_{DS,On}$ and BVDSS depending on the material properties is derived in [24]. It is given by Eq. 2.1 for vertical and by Eq. 2.2 for lateral power devices. The denominator $\epsilon_r \cdot \mu_n \cdot E_{Cr}^3$ is also known as Baliga's figure-of-merit (BFOM) [25] and consists only of technology parameters of the respective material.

$$R_{DS,On\_vertical} \cdot A = \frac{4 \cdot BVDSS^2}{\epsilon_0 \cdot \epsilon_r \cdot \mu_n \cdot E_{Cr}^3} \tag{2.1}$$

$$R_{DS,On\_lateral} \cdot A = \frac{BVDSS^2}{\epsilon_0 \cdot \epsilon_r \cdot \mu_n \cdot E_{Cr}^3} \tag{2.2}$$

By inserting the material characteristics from Table 2.1 into Eq. 2.2, the theoretical limit for the specific $R_{DS,On}$ at a given breakdown voltage BVDSS of the respective technology can be calculated. The highest denominator is achieved with the values of GaN technology. Thus, a transistor with a defined $R_{DS,On}$ and BVDSS required less area if fabricated using GaN technology than for silicon or SiC technologies. However, due to various imperfections of the process technology and the fabricated wafers (see Sect. 2.2), these theoretical limits are typically not achieved for real devices.

Table 2.1 Comparison of fundamental technology properties

	Unit	Silicon	SiC	GaN
E_G	eV	1.12 [19]	3.26 [19]	3.4 [1]
E_{Cr}	MV/cm	0.3 [20]	2.4 [20]	3.3 [19]
μ_n	$cm^2/(Vs)$	1400 [19]	950 [19]	2000 [21]
v_{sat}	10^7 cm/s	1.0 [20]	2.0 [20]	3.1 [22]
ϵ_r		12	9.8	9.5–10.4
BFOM	GW/cm^2	0.04	11.4	60.5

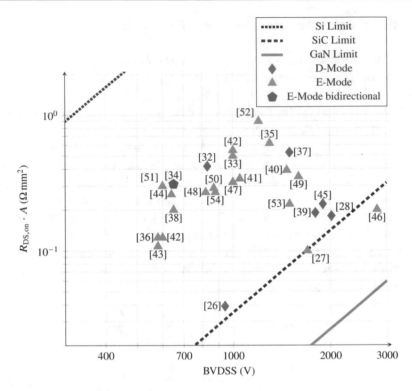

Fig. 2.1 Performance of published GaN transistors with respect to theoretical limits

Figure 2.1 depicts the specific on-resistance versus breakdown voltage for various published GaN transistors. The theoretical limits for Si, SiC, and GaN are calculated based on their respective BFOM and added to the plot. This representation is well established in both primary [26–28] and secondary [29–31] literature to compare different published transistors and technologies with respect to each other and the theoretical limits.

All of the GaN transistor represented in Fig. 2.1, already from early works [44, 45], exceed the theoretical Si-limit. More recent works [46] even surpass the theoretical limit for SiC. Thus, GaN is not only theoretically a promising technology for compact high-voltage transistors, but it is proven that GaN can exceed the performance of Si and SiC power transistors.

Despite the higher cost for processed GaN wafers, the smaller area utilization of GaN transistors is beneficial to keep the cost per die within affordable limits. Some predictions expect GaN to achieve price parity with silicon not only on system level but also on device level [55]. Furthermore, the smaller area for same $R_{DS,On}$ and BVDSS also leads to smaller specific capacitances for GaN transistors. Hence, faster and more efficient switching is possible. The comparison of commercially available silicon [56, 57] and GaN [58, 59] transistors with similar $R_{DS,On}$ and BVDSS shows that GaN transistors typically require a 20–50 times lower gate charge (Q_g) for turning on and have a two to three times lower energy-related output capacitance ($C_{O(ER)}$) (see also Table 2.2). The absence of any reverse recovery in the power path of GaN transistors leading to almost zero Q_{rr} is most notable (see Sect. 2.3). Especially the lower $C_{O(ER)}$ and the zero reverse-recovery charge (Q_{rr}) are valuable properties of GaN transistors, since the recharging of these capacitances is the dominant loss mechanism in hard-switched high-voltage applications. Summarizing, the combination of small size and high performance is the reason for increasing interest in GaN transistors for compact, high-voltage power supply applications.

GaN Technologies with Enhancement-Mode Characteristic

The growth of GaN as a single crystal to obtain bulk GaN wafers is challenging [9] and thereby expensive. Nevertheless, GaN transistors on bulk GaN substrates have been demonstrated [53, 54]. However, for cost-sensitive mass-market applications, GaN is usually grown on foreign substrates, where silicon is the most available and least expensive choice. However, growth of a GaN epitaxial layer on silicon wafers comes with its own set of challenges, i.e. the crystals have different lattice constants and different thermal expansion coefficients. To avoid wafer bending and even shattering during the fabrication process, a multi-layer buffer is grown on top of a silicon carrier wafer to compensate for mechanical stress. Nevertheless, the growth of the GaN layer is not yet developed to perfection and it shows a higher defect density than bulk silicon and SiC technologies. Thereby, the performance of GaN transistors is still below their theoretical limit and only few examples have passed the SiC limit, yet (see Fig. 2.1).

For the application of GaN transistors in power electronics, fail-open enhancement-mode (e-mode) characteristic is desired to ensure safe operation. In the natural GaN device, depicted in Fig. 2.2a, electrons are generated as carriers by spontaneous piezoelectric polarization at the AlGaN/GaN interface without the need of an external control voltage [60]. These natural devices show a depletion-mode (d-mode) characteristic and do not support fail-open operation.

Different methods are reported in literature in order to achieve an e-mode characteristic. One solution is to use a HV d-mode GaN transistor as cascode device together with a low-voltage (LV) e-mode silicon transistor, Fig. 2.2b. Such configurations are commercially available (e.g. [61]) and offer not only the power switching device but also additional features such as the gate driver and protection circuits (see Sect. 2.5). However, the performance of this method is limited due to the series connection of two transistors in the power path. Additionally, the multi-chip solution increases the complexity for assembly and test.

GaN transistors with inherent e-mode characteristic achieving a sufficiently high threshold voltage of around 1 V are first presented in [62]. There, fluorine ions are implanted in the AlGaN layer, as illustrated in Fig. 2.2c. The negatively charged ions deplete the 2DEG channel below at zero gate-source voltage (V_{GS}) and thereby shift the threshold voltage

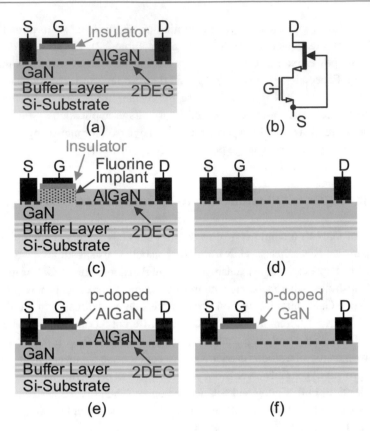

Fig. 2.2 Illustrated cross sections of **a** natural d-mode, **b** d-mode cascode configuration [61], **c** fluorine implant e-mode [62], **d** recessed-gate e-mode [51], **e** p-AlGaN gate e-mode [44], and **f** p-GaN gate e-mode [50] GaN transistors

(V_{th}) of the transistors to positive values. If a positive $V_{GS} > V_{th}$ is applied, the electrons are re-accumulated and the device is turned on. Fluorine-based e-mode technology has been an academic research topic over some years for stand-alone power devices [63, 64] as well as integration of circuits in GaN [65, 66]. There has also been some industrial research on this method [52]. However, the interest in fluorine doping faded as other e-mode technologies for GaN appeared and already one year later in 2012, the same authors published e-mode GaN devices using gate-recess technology [67] as depicted in Fig. 2.2d.

The recessed-gate technology was first presented by [51] in 2006. While an e-mode characteristic can be achieved by keeping the AlGaN layer very thin over the full transistor length [2], the device performance is reduced due to lower carrier density. In [51], the AlGaN is deposited as a thick layer. In a next step, it is etched to a very thin layer only in the gate region. Thus, an e-mode characteristic is achieved while the good performance of the 2DEG channel below a thick AlGaN layer is maintained in the drain drift region. As an industry company, NEC published a method for manufacturing a recessed-gate e-mode transistor in 2009 [33].

The most promising e-mode technologies today use a p-doped layer between the gate metal and the AlGaN layer to affect the conduction bands of the hetero junction, Fig. 2.2e, f. The earlier of these two approaches is presented in [44] using p-doped AlGaN forming a gate injection transistor (GiT). This type of GaN transistor requires a constant gate current in the mA-range to be fully turned on. Thereby, they achieve low $R_{DS,On}$. Additionally, the gate current leads to an intrinsic clamping capability of the transistor gate making it robust against transient V_{GS} overshoots. However, the DC gate current requirement poses some challenges for the gate driver. Typically, high-side drivers with a floating reference potential are supplied utilizing a bootstrap technique, where the supply voltage is stored on a capacitor. To provide a DC gate current for the transistor, the size of the bootstrap capacitor has to be significantly larger than for the case, when only the transient turn-on charge has to be provided as it is common for silicon field-effect transistors such as super junction FETs. Nevertheless, gate injection transistors have been developed until productization and are commercially available [58].

In order to reduce the DC gate current of the transistor, a p-doped GaN layer is introduced replacing the p-doped AlGaN [50]. Together with a tungsten-based gate metal, the p-GaN layer forms a Schottky junction in reverse bias, significantly reducing the gate current [68]. While this basically eliminates the DC gate current for keeping the device turned on, it also removes the clamping capability of the gate itself. Thus, the gate structure is sensitive to overvoltages, which may result in rapid aging or catastrophic failure of the transistor [69, 70]. In contrast to gate drivers for GiTs, a bootstrap supply for Schottky gate p-GaN transistors only needs to supply the turn-on gate charge and can therefore be sized much smaller. However, the driver has to be designed carefully in order to avoid gate-source voltage overshoots. This is discussed in detail in Sect. 2.4. Discrete e-mode GaN transistors using p-GaN gate are commercially available from multiple suppliers [71, 72]. The p-GaN technology is also available as a foundry process for academic and industrial purposes [41].

While state-of-the-art high-voltage silicon superjunction [73] and SiC [74] transistors are vertical devices, the high-voltage transistors in GaN technology are formed by a lateral structure as illustrated in Fig. 2.2. Thereby, monolithic integration is possible, as demonstrated in [75–77] and other publications. Figure 2.3 shows the illustrated cross sections of devices, which can be fabricated utilizing a high-voltage p-GaN gate e-mode process flow without any additional masks or process steps [78–80].

The high-voltage transistor is similar to the structure shown in Fig. 2.2f. By removing the drain extension, a more compact low-voltage transistor can be formed, which achieves lower specific on-resistance and capacitances due to the shorter device length. When the gate stack is skipped, the naturally present 2DEG at the AlGaN/GaN interface can be used to form a resistor. Additionally, the available metal layers can be employed to form metal insulation metal (MIM) capacitors. These devices can be fabricated on one die and enable integration of different functions. For the IC designs presented in this work such a p-GaN

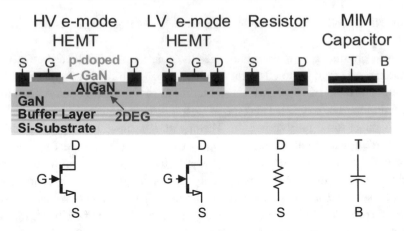

Fig. 2.3 Illustrated cross section of devices available in a standard p-GaN gate e-mode process

gate ("p-doped GaN" in Fig. 2.2f) GaN-on-Si technology is used. An overview of different integration levels achieved in GaN technology is provided in Sect. 2.5.

Carrier Mobility in the Two-Dimensional Electron Gas

In GaN technology, the conductive channel for both transistors and resistors is formed by a lateral 2DEG (see Fig. 2.3). Hence, the basic properties of the 2DEG determine the characteristics and performance of both devices central for the integration of circuits using this fabrication process. As indicated by the name, the performance of AlGaN/GaN HEMTs is related to the high mobility μ_n of electrons at the AlGaN/GaN interface (see Table 2.1). This part examines some properties of the 2DEG in GaN transistors, especially the temperature dependency in comparison to silicon technologies.

The carrier mobility is generally defined as the ratio between the carrier drift velocity and the electrical field accelerating the carriers as given in Eq. 2.3.

$$\mu_n = \frac{v_{drift}}{E} \tag{2.3}$$

At room temperature, μ_n is typically around 2000 cm^2/(V s) for GaN (from Table 2.1). The basic drain current equation for a field-effect transistor in linear region with small V_{DS} (Eq. 2.4) shows a linear relation between current of the transistor and the carrier mobility.

$$I_D = \mu_n \cdot C_{ox} \cdot \frac{W}{L} \cdot \left(V_{GS} - V_{th} - \frac{V_{DS}}{2} \right) \cdot V_{DS} \tag{2.4}$$

Thus, the temperature coefficient of the mobility has a direct influence on the drain current of the transistor and, accordingly, on the 2DEG resistor. Literature reports a strong temperature coefficient of the electron mobility in GaN transistors. From measurement plots in [81], a temperature coefficient of -5,700 ppm/K can be estimated for the mobility in AlGaN/GaN hetero structures. Reference [82] investigated the power law for the temperature dependence of the mobility according to Eq. 2.5 where A is a constant, T_{abs} is the absolute temperature, and γ is the power coefficient. The power law is based on the assumption that the mobility in GaN is limited by phonon scattering at temperatures above -80°C.

$$\mu_n = A \cdot T_{abs}^{-\gamma} \tag{2.5}$$

For $\gamma = 2.8$ [82], the temperature coefficient of the mobility can be calculated to be $-5,500$ ppm/K between room temperature of 27°C and 125°C. This is in accordance with the value estimated from characterization plots in [81]. In silicon, $\gamma \sim 2.5$ [83] which leads to a similar temperature coefficient of $-5,100$ ppm/K.

The transistor on-resistance $R_{DS,On}$ can be calculated as ratio between the drain-source voltage V_{DS} and the drain current I_D. Inserting the relation from Eq. 2.4, this results in Eq. 2.6 for the $R_{DS,On}$ in linear region, where $R_{DS,On} \propto 1/\mu_n$.

$$R_{DS,On} = \frac{V_{DS}}{I_D} = \frac{1}{\mu_n \cdot C_{ox} \cdot \frac{W}{L} \cdot \left(V_{GS} - V_{th} - \frac{V_{DS}}{2}\right)} \tag{2.6}$$

The mobility with its power law for the temperature dependence appears in the denominator. Consequently, a similar power law can be derived for the on-resistance of a transistor, where the term $T_{abs}^{-\gamma}$ of the mobility (from Eq. 2.5) can be moved to the nominator by swapping the sign of the power coefficient γ. This results in Eq. 2.7, where B contains everything from Eq. 2.6 except for the temperature coefficient of μ_n.

$$R_{DS,On} = B \cdot T_{abs}^{\gamma} \tag{2.7}$$

With $\gamma = 2.8$ for GaN the temperature coefficient for $R_{DS,On}$ can be calculated to be 12,400 ppm/K. For silicon with $\gamma = 2.5$, it is 10,500 ppm/K, respectively. This is confirmed by $R_{DS,On}$ values at different temperatures of commercially available power transistors. Reference [72] reports a resistance change from 50 mΩ to 129 mΩ for a GaN transistor, when the temperature increases from 25°C to 150°C. The related temperature coefficient for the $R_{DS,On}$ can thereby be calculated to be 12,600 ppm/K. From the data provided for silicon [84], the temperature coefficient for $R_{DS,On}$ is 10,100 ppm/K.

The temperature coefficient of the 2DEG resistance strongly influences the performance of integrated circuits in GaN. As an example, the propagation delay of most integrated circuit is composed of distributed RC time constants. When each 2DEG resistance value increases by a factor of two at a temperature rise from room temperature to some $100°C$, also the propagation delay of circuits containing resistors or transistors increases by the same factor of two. The implications of this temperature dependence on different circuits are discussed in various sections of this work, e.g. in Sects. 3.1, 3.2, and 4.2.

2.2   Technology Challenges in GaN

In comparison with silicon, today's e-mode GaN process technology is less mature. Especially the growth of GaN epitaxial layers on affordable silicon wafers shows imperfections. While the buffer layer (see Fig. 2.2) is engineered to reduce lattice mismatch and mechanical stress caused by mismatched thermal extension to a minimum, the GaN layer still shows higher crystal defect densities than common for today's silicon wafers. The crystal defects as well as the buffer structure and, additionally, the junction gate create various possibilities for charge trapping, which in turn affects the threshold voltage, the transconductance as well as the current-carrying capability of the transistors.

A particular challenge is related to the large difference between electron and hole mobility in GaN, which leads to the absence of suitable complementary transistors. A brief overview of various effects and published countermeasures is provided in this section with a focus on the implications for circuit design in GaN.

P-Type Devices in GaN

The developments in silicon technologies show that the availability of complementary devices boosted the performance of integrated circuits dramatically. On one hand, it allows for integration of large and complex digital circuits in an efficient manner by avoiding static current consumption. On the other hand, it also enables analog CMOS design techniques with various breakthroughs in terms of accuracy, achievable gain, common-mode range, and lower minimum supply voltage requirements.

In general, both carrier types, electrons and holes, also exist in GaN technology. Hence, the fabrication of complementary n-type and p-type transistors is possible. In silicon technology, the mobility of holes is about three times lower than the mobility of electrons. Thus, the PMOS pull-up transistor of a silicon CMOS logic inverter is typically designed with three times the channel width of the pull-down NMOS in order to achieve a symmetrical behavior in terms of pull-up and pull-down strength. In GaN, however, the factor between hole and electron mobility is not three, but somewhere between 50 and 100 [85]. According to Eq. 2.4, the drain current density I_D/W of a transistor operated in linear region is directly proportional to the carrier mobility μ. Reference [86] reports a maximum on-state drain current density for a p-type transistor of less than 1 mA/mm, [87] reports 6.1 mA/mm. In contrast to that, state-of-the-art n-type GaN transistors with similar drain-source voltage rating show on-state currents well above 300 mA/mm [66]. Despite the poor performance of p-type GaN devices, academic research demonstrated the possibility of complementary logic circuits in GaN [85, 88–90]. Figure 2.4 shows the cross section of both, n-type and p-type, transistors fabricated with one monolithic process flow. For the p-type device, additional process steps such as the growth of a thick p-GaN layer used for the conductive channel as well as recess etching and passivation of the gate area are required. Moreover, an additional metal is necessary to form ohmic drain and source contacts to the p-GaN layer. The p-ohmic metal in Fig. 2.4 is one alloy which leads to a ohmic contact to the p-GaN channel of the transistor. This is in contrast to the tungsten-based gate metal, which leads to a rectifying Schottky contact to the p-GaN layer in the gate stack. The n-ohmic metal is yet another metal required to form an ohmic contact for the electrons of the 2DEG channel of the n-type GaN transistors.

While the basic functionality is proven, the large asymmetry between n-type and p-type devices limit the performance and applicability of complementary logic. Reference [88] reports a specific on-resistance $R_{DS,On}$ of 10Ω mm for the n-type transistor versus 1300Ω mm for the p-type device. This leads to a large difference between rise and fall time of the implemented inverter with 90 ns and 670 ns, respectively, despite a ten times larger width is used for the p-type transistor. A more recent publication [90] shows similar resistance values for the transistors with 9 Ω mm for a n-type transistor and 1,200 Ω mm for a p-type device, respectively. Due to a suitable inverter threshold voltage as well as lower drain-source leakage for the p-type device, an average inverter delay of 50 ns is reported as a significant improvement. However, compared to CMOS inverters in today's silicon technologies, this is still slow (see Sect. 3.2). This is caused by the relatively large $R_{DS,On}$ of the p-type devices with respect to the capacitances that have to be recharged when the output state of an inverter changes. The GaN roadmap published in [91] expects a performance improvement by a factor of five for p-type GaN transistors in the near future. While this would reduce the asymmetry between the complementary devices, it still leaves a factor of 25 between the specific on-resistances and, consequently, between the switching transition delays.

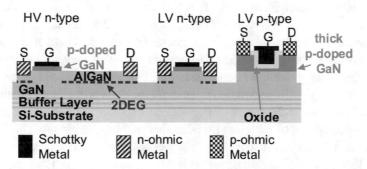

Fig. 2.4 Illustrated cross section of complementary transistors in monolithic GaN process flow [90]

While complementary GaN transistors are desirable and would enable a much higher performance for integrated GaN circuits, the present and expected near-term performance of p-type GaN transistors is very limited. Considering the additional masks and fabrication steps for p-type devices, it is not surprising that the interest in complementary GaN is limited to academic research while commercially available GaN technologies offer only n-type transistors. Thus, this work evaluates the performance of resistor-transistor logic (RTL) in a state-of-the-art n-type GaN process in Sect. 3.2 and investigates circuit design techniques to improve the performance of analog (Sect. 3.1) and mixed-signal (Sect. 3.3) circuits without using p-type devices. Thereby, the methods and results of this work can be transferred to most e-mode GaN technologies and do not rely on rarely available devices.

Charge Trapping and Related Effects

When GaN transistors were initially applied for high-voltage switching operations in research, it was soon discovered that the performance of the transistors significantly degraded after both on- and off-state stress. Reference [92] provides an early investigation of the phenomena called current collapse and dynamic $R_{DS,On}$. There, it is demonstrated that the achievable drain current of a GaN transistor reduces significantly after high-voltage off-state stress. This phenomenon is called current collapse. When the transistor is turned on for some period of time after the off-state stress, the device may be able to recover from the reduced current capability. Together with threshold voltage instabilities caused by on-state stress at the gate, this leads to a dynamic $R_{DS,On}$ behavior, which shows time-dependent changes of on-resistance and the associated drain current while the transistor is turned on [93, 94] showed that hot electrons caused by high-voltage off-state stress have a major contribution to the current collapse. Figure 2.5a illustrates a fundamental high-voltage GaN transistor with an extended drain drift region. When the transistor is off, a large peak electrical field is present at the gate edge towards the conductive 2DEG of the drain drift region. This may lead to electrons with a high energy and thus to charge trapping. As one countermeasure, a source-connected field plate is proposed as depicted in Fig. 2.5 (b). The field plate basically acts as an additional high-voltage gate and distributes the electrical field over a larger area in the 2DEG at the AlGaN/GaN interface. Thereby, the field plate prevents hot electrons from getting trapped at the gate edge. In contrast to that, [95] shows that field plates can lead to trapped electrons in the SiN passivation layer above the drain drift region, which may result in a full current collapse. However, by deposition of appropriate passivation layers, these trapping mechanism can be mitigated [95]. Another technique is proposed in [96], where a so-called hybrid drain is formed by additional p-doped GaN in the drain access region, see Fig. 2.5c. This p-GaN injects holes into the transistor channel, which can recombine with trapped electrons and thereby reduce the current collapse to a minimum.

While some fabrication techniques achieved acceptably low current collapse and small dynamic $R_{DS,On}$ behavior, the investigation of charge trapping as root cause is an ongoing

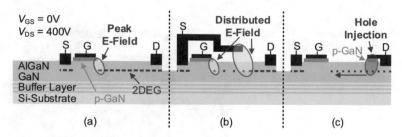

Fig. 2.5 Illustrated cross section of **a** basic high-voltage e-mode HEMT, **b** high-voltage e-mode HEMT with source-connected field plate [94], **c** high-voltage e-mode HEMT with additional p-GaN as hybrid drain [96]

research topic. In the GaN transistor, there are various locations, where charge trapping can occur. References [97] and [98] show different charge trapping mechanisms in the gate region, where traps for both, electrons and holes, exist inside of the p-GaN layer and also at the interface between p-GaN and AlGaN. This is illustrated in Fig. 2.6 at locations (a) and (b). When electrons are trapped in the gate stack, they repel the electrons in the channel below and thereby reduce the 2DEG density. This leads to higher $R_{DS,On}$ of the transistor and can also be seen as a higher threshold voltage for the device. During on-state stress, electrons from the channel can accumulate in the p-GaN leading to a positive drift of the threshold voltage, which is called positive bias temperature instability (PBTI). Reference [99] shows that a high enough gate voltage can inject holes from the p-GaN layer in the gate. Thereby, trapped electrons can be neutralized. However, additional holes can then be trapped at the p-GaN/AlGaN interface. This, in turn, accumulates more electrons in the channel below, strengthening the 2DEG and effectively reducing the threshold voltage of the transistor. In accordance with this, [100] demonstrates that after a gate stress with $V_{GS} = 6$ V, the threshold voltage is higher than after a gate stress at $V_{GS} = 4$ V. However, when the stress voltage is increased to $V_{GS} = 9$ V, the threshold voltage after stress is even lower than for a gate stress of $V_{GS} = 4$ V. In a similar mechanism, holes injected by the p-GaN as part of the hybrid drain (Fig. 2.5c) may also be able to recombine with trapped electrons, neutralizing PBTI effects. Some of these trapping effect leading to variations of the threshold voltage may contribute to the elevated low-frequency noise of GaN transistors presented in Sect. 3.1.

Hole trapping at the interface between the 2DEG channel and the AlGaN back barrier as part of the buffer layer between silicon carrier wafer and GaN epitaxial layer is described in [101]. This is indicated as location (c) in Fig. 2.6. Trapped charges in that location effectively act as a back gate, which also modulates the transistor channel. When the gate contact is formed by an ohmic metal, a negative threshold voltage shift is observed under positive V_{GS} stress. In contrary, the same stress conditions lead to a positive threshold voltage shift when the gate metal forms a Schottky junction in reverse bias together with the p-GaN in the gate stack. Accordingly, [102] describes trapping in the buffer layer of a d-mode GaN

Fig. 2.6 Illustrated cross section of a high-voltage e-mode HEMT with source-connected field plate depicting various trap locations

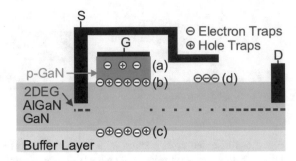

MISHEMT leading to a V_{th} shift by -1.8 V after $V_{DS} = 400$ V off-state stress. There, 70% of the threshold voltage shift is recovered within 30 ps after the transistor is turned on. The electron traps at the AlGaN surface towards the passivation layer (location (d) in Fig. 2.6) are described in [95]. These traps can be eliminated by proper material selection for the passivation layer above the drain drift region.

The various trapping effects may not only occur in high-voltage power transistors, but also in low-voltage devices utilized for integrating circuits in GaN. Since the gate structure and voltages are typically the same for both the high-voltage and the low-voltage transistor (see Fig. 2.3), all gate-bias-related effects also occur in low-voltage transistors. For analog circuits, where transistors usually are operated in saturation region with gate-source voltages around the threshold voltage, the carrier density is low leading to fast electrons. Therefore, also trapping effects caused by these hot electrons may apply to low-voltage transistors when operated under certain conditions.

The investigation of various trapping effects in high-voltage GaN transistors, which is the driving component for the GaN process development, is still ongoing. Further research in this area, also extended to trapping and degradation effects in low-voltage transistors, is required. Especially when higher levels of integration in GaN are desired, the mechanisms and implications on circuit design have to be understood. Some considerations based on analog characterization results are provided in Sect. 3.1.

Random Crystal Defects in GaN and Effects on Devices

Another challenge in GaN technology is related to the crystal growth of GaN, especially on foreign substrates such as silicon wafers. Reference [103] shows a reduction of the carrier mobility in the 2DEG when the density of a certain crystal defect, called threading dislocation, increases. The reported mobility reduces by up to 16.8 % when the defect density is at 10^{10}cm^{-2}. This is several orders of magnitude larger than the threading dislocation densities in silicon technology, which is reported to be between 10^4cm^{-2} and 10^5cm^{-2}. [104, 105] examine the effects of dislocations on the transistor parameters transconductance g_m and drain current I_D by computer simulation. A degradation of the drain current by 12%

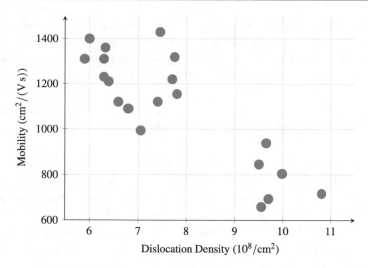

Fig. 2.7 Characterized electron mobility vs. dislocation defect density from [106]

is reported, when the dislocation density increases from 10^8cm^{-2} to 10^{10}cm^{-2}. A similar reduction is estimated for g_m. Based on the fundamental equations for field-effect transistors (see Eq. 2.4), both g_m and I_D show a linear dependency on the carrier mobility. Hence, the reduced mobility reported in [103] matches well with the reduced drain current and transconductance presented in [105]. Furthermore, [105] predicts a drain current reduction of 53% when the dislocation density increases to 10^{11}cm^{-2} and even a reduction by more than 90% for a dislocation density of 10^{12}cm^{-2}.

A characterization of defect density and electron mobility for AlGaN/GaN sample structures fabricated on a SiC carrier wafer is provided in [106]. The results show a reduction of the electron mobility by 50% when the dislocation density increases from $6 \times 10^8 \text{cm}^{-2}$ to $11 \times 10^8 \text{cm}^{-2}$. This is illustrated in Fig. 2.7.

Furthermore, it is demonstrated that the mobility can vary by more than 30% between adjacent sample structures with a dimension of $40 \times 40 \mu \text{m}$ [106]. Hence, the dislocations seem to be spread randomly across the AlGaN/GaN interface. Additionally, [107] proves that magnesium used as dopant for p-GaN layer in e-mode transistors leads to additional threading dislocation defects.

In the context of circuit integration in GaN, this implies serious challenges for absolute accuracy of device parameters. Furthermore, the mismatch of devices based on the 2DEG is expected to be considerably higher than in established silicon technologies. Typically sized low-voltage GaN transistors utilized for analog and digital integrated circuits may cover an AlGaN/GaN area from some 5 μm^2 up to 500 μm^2. This is smaller than the $40 \times 40 \mu \text{m}$ reported in [106]. Hence, potential averaging effects are even smaller for the transistor sizes used in integrated circuits. Together with the additional dislocations caused by the p-GaN gate (see Fig. 2.3), matching of integrated GaN transistors can be expected to be a serious challenge. Similar to that, also the matching of 2DEG resistors might be challenging. However, due to the absence of p-GaN in the resistor structure, the matching is most likely better than for transistors. An investigation and characterization of the matching properties of GaN transistors and resistors is provided in Sect. 3.1.

2.3   Benefits of GaN for Integrated Power Electronics

As a wide bandgap material, GaN shows several benefits for the performance of high-voltage transistors (see Sect. 2.1). In addition to that, there are several more advantages for the implementation of integrated circuits in GaN related to the device structure of components (see Fig. 2.3). They include naturally isolated devices as well as the absence of junctions in the transistor channel. This section discusses the technology-related benefits for integrating analog, digital, and mixed-signal circuits in a power process. They extend to parasitic bipolar devices; isolation structures; and, related to that, the integration of different high-voltage functions.

Power Process Without PN-Junctions

The absence of pn-junctions in the power path is a major benefit of GaN technology for power electronics. The GaN HEMT is formed by a hetero-structure as illustrated in Fig. 2.8a. The only junction in the transistor is located at the gate stack. They are formed by a Schottky junction in reverse bias at the intersection of the gate metal and the p-GaN in series with a back-to-back configuration of a p-i-n diode from the p-GaN to the 2DEG [108, 109]. However, GaN transistors have no bipolar junctions in the power path and there exists no substrate diode.

In contrast to that, silicon MOSFETs are formed with multiple pn-junctions leading to parasitic diodes and even parasitic bipolar transistors. For comparison, a cross section of a n-type lateral drain-extended metal–oxide–semiconductor (LDMOS) is in silicon technology as depicted in Fig. 2.8b. The various pn-junctions can provide unwanted current paths in inductive switching applications typical for power converters. Furthermore, even the parasitic PNP-bipolar transistor with the backgate (B) connection as emitter, the drain (D) connection as base, and the substrate as collector may turn on. Thereby, the switching losses of the transistor may increase significantly and the current flow can even lead to destruction of the device, as has been discussed in [110–112]. If the backgate (B) is not shorted to the source, also the parasitic NPN transistor with drain as collector, backgate as base and source as emitter may turn on in some conditions. In order to avoid unwanted current flow with potential harm to the device, all areas forming the various junctions in the power path of the silicon NMOS transistor have to be carefully biased in all operating

Fig. 2.8 Illustrated cross section of **a** a GaN HEMT, **b** a silicon LDMOS with parasitic diodes

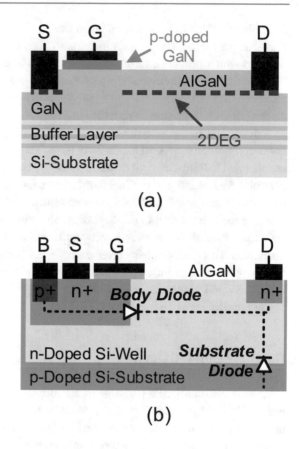

(a)

(b)

conditions. These considerations are not required for GaN transistors and thus simplify the design of circuits in GaN technology.

However, in various applications the body diode of the silicon transistor is used as free-wheeling diode to provide a path for inductive currents during switching transitions. Using the body diode works very well for soft switching applications when the switching happens at either zero voltage or zero current. Also for hard switching applications this technique is common. Nevertheless, the dissipation of carriers required to form the current block-ing depletion region (reverse recovery) dramatically increases switching losses of silicon transistors in hard switching operation.

In contrast to that, GaN HEMTs do not possess a pn-junction forming a body diode. However, they allow a current flow in both directions, from drain to source and also in the opposite direction. This can be exploited to achieve a behavior of GaN transistors similar to the body diode conduction of silicon devices. The operation mode when $V_{GS} < V_{th}$ and $V_{DS} < -V_{th}$ is referred to as quasi-body-diode or third-quadrant conduction. The voltage drop across the quasi-body-diode with typical 3 to 5 V is considerably higher than the nominal forward voltage of 0.7 V for a pn-junction in silicon. This leads to increased conduction losses in third-quadrant operation when the GaN transistor is not actively turned on. The third-quadrant conduction losses increase to even higher levels when bipolar gate drivers are used which provide a negative gate-source voltage for turning the GaN transistor off. Thus, multi-level gate drivers are proposed for e-mode GaN transistors [113] which have also been developed for a fully integrated gate driver IC [114]. A more detailed discussion on driving GaN transistors is provided in Sect. 2.4.

Independent of the gate driver topology, the voltage drop across the GaN transistor in quasi-body-diode operation is relatively large. Thus, the time period when GaN transistors are operated in third-quadrant conduction should be kept to a minimum. Hence, a dead-time control is an integral part of GaN half-bridge controller and driver ICs in order to achieve high power conversion efficiency, as is presented in [115] and others.

Since the current path in third-quadrant operation is not provided by a bipolar junction, GaN transistors show no reverse-recovery losses while supporting all required conduction operations. This absence of any reverse recovery is one key benefit of GaN transistors utilized for hard switching of inductive loads. Thereby, high-frequency operation can be achieved using GaN and topologies such as the totem pole PFC converter are enabled, achieving power efficiencies as high as 99% [116, 117].

Naturally Isolated Devices

Integrated silicon MOSFET transistors are formed by multiple p-doped and n-doped wells, one inside the other (see Fig. 2.8b). In order to achieve a compact layout with low area utilization, these wells can be shared between multiple transistors. In standard bulk silicon process technologies, all transistors share a usually highly p-doped, low resistive substrate. Depending on the particular transistors and their functions, it is also common to share the n-doped well and even the p-body of multiple NMOS transistors, see Fig. 2.8b. This also applies to PMOS transistors, where the p-doped well and the n-body may be shared accordingly for multiple transistors. Thus, careful considerations are required to select the proper biasing for all wells in order to maintain an isolating junction in reverse bias between different devices under all operation conditions. On top of that, some transistors, especially ones with higher voltage ratings above 20 V, require additional isolation guard rings composed of various junctions. On one hand, this adds complexity for the biasing of the junctions and, on the other hand, the guard rings around the transistors require a significant additional area.

Modern process technologies offer trench isolation, where a cavity is etched into the conductively doped silicon between different devices which should be isolated from each other. The cavity is filled with isolating material such as silicon oxide. Thereby, isolation is achieved without requiring any junctions and the associated biasing concept. Other processes use silicon-on-insulator (SOI) techniques, where the silicon base wafer is only a mechanical carrier. That silicon wafer is completely covered with an insulator (i.e. silicon oxide) and both n- and p-doped silicon layers are selectively grown on the insulator. While this provides good isolation capabilities, the fabrication involves many additional process steps and a considerably higher number of masks is required.

GaN technology is somewhat similar to SOI technologies since GaN itself is a good insulator. In the cross section of Fig. 2.8 (a), the "Si-Substrate" corresponds to the mechanical carrier wafer of an SOI technology and the "Buffer Layer" as well as the "GaN" layer has the same electrical role as the insulator. The conductive 2DEG layer is only formed at the heterojunction of GaN and the selectively grown AlGaN. Hence, all devices in GaN technology are naturally isolated as long as the AlGaN layer is not connected between the devices (see Fig. 2.3). This is a major advantage over bulk silicon technologies, since no additional isolation structures are required. Moreover, careful biasing considerations are not needed, since no conductive wells exist.

The critical electrical breakdown field of GaN is above 3 MV/cm (see Table 2.1). Thus, voltages up to 300 V can be isolated with 1 μm spacing in GaN. For reliability purposes, this distance is usually extended to reduce the maximum electrical field strength. However, multiple isolated high-voltage devices can be integrated on one GaN die with reasonably low area overhead. Close-proximity sensing of high-voltage signals is one application for high-voltage devices integrated together with the power transistor (see Sect. 4.3). Another valuable application field for high-voltage integration is the implementation of high-voltage startup and supply voltage regulator in order to reduce the system complexity on PCB (discussed in Sect. 4.4).

The silicon carrier wafer in GaN-on-Si technology is typically highly doped and connected to ground via a thermal pad common of power transistor packages. Hence, the silicon substrate can act as a back gate for GaN transistors and influence the resistance and drain current of the channel. Nevertheless, the distance between silicon and the 2DEG channel is typically larger than 10μm. Thus the backgating effect can be neglected for low-voltage applications such as integrated analog and digital circuits with voltage levels typically well below 10 V [118]. For voltages around 100 V and higher, the backgating effect of the silicon substrate cannot be neglected anymore [119].

Figure 2.9 shows an exemplary configuration for a monolithic GaN half-bridge. In the depicted case, the drain of the high-side transistor Q_{HS} is connected to a typical DC voltage rail $V_{bus} = 400$ V. When the low-side transistor is turned off and the high-side transistor is

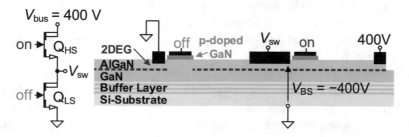

Fig. 2.9 Schematic and illustrated cross section of a monolithic integrated GaN half-bridge

turned on, also the switching node V_{sw} and the 2DEG channel of Q_{HS} are at 400 V. With the silicon carrier wafer grounded, this results to a back gate voltage $V_{BS} = -400$ V between silicon and 2DEG. Such a large back gate voltage significantly weakens the 2DEG channel and may even fully deplete it. Reference [119] shows a drain current reduction by 50% for a back gate voltage of -200 V. Hence, a substrate isolation in this GaN-on-SOI process flow is proposed in order to individually bias the back gate for each high-voltage power transistor. This enables the monolithic integration of a 200 V GaN half-bridge on one die. However, higher voltages required for applications operated directly at the 230 V power grid are not yet supported by the GaN-on-SOI technology in [119]. The voltage capability is likely limited by vertical breakdown of the transistor. A review on different integration levels around GaN technology including package integration of a 650 V half-bridge is provided in Sect. 2.5.

2.4   Gate Loop Challenge for GaN Drivers

Due to low specific capacitances, GaN transistors support higher switching frequencies and faster slew rates than comparable silicon transistors. High-frequency operation is desired to reduce the inductance and capacitance values and the associated physical size of required passive energy storage components. Thus, the power density can be significantly increased (see also Sect. 5.4). Fast slew rates are desired in order to minimize switching losses, especially I-V overlap losses in hard-switched applications. However, at higher switching frequencies and faster slew rates, parasitic inductances and capacitances in the gate loop have a stronger influence and can lead to significant ringing as well as harmful voltage overshoots. This is an even more severe challenge for GaN transistors, since they show much smaller threshold voltages and maximum gate-source voltage ratings than their silicon and SiC counterparts. Thus, this section examines the challenges for GaN drivers related to the parasitics in the gate loop.

Table 2.2 provides a comparison of critical parameters for high-voltage power transistors. It shows typical transistors in different semiconductor technologies suitable for offline power converters with a breakdown voltage BVDSS of 650 V and similar current-carrying capabilities of around 30 A. The critical parameters for switching applications are the threshold voltage V_{th} as well as the gate-source charge for turn-on $Q_{gs,on}$. Also the ratio between $Q_{gs,on}$ and the charge coupled to the gate through the Miller capacitance $Q_{gd,400V}$ during switching transitions is critical to ensure safe off-state behavior. The GaN transistor shows the lowest values for these parameters not only indicating its superior switching performance but also the sensitivity towards ringing.

In this section, the challenges for driving GaN transistors due to tight gate voltage requirements in combination with high switching speed are investigated. Requirements for the gate loop inductance for both, safe turn-on without harmful gate voltage overshoots and safe off-state during slewing without potentially destructive cross currents, are investigated. Different techniques to ensure proper operation are presented and discussed.

Table 2.2 Comparison of 600 V discrete power transistors

	GaN HEMT [72]	Silicon superjunction [84]	SiC MOSFET [120]
BVDSS	650 V	600 V	650 V
$R_{DS,On}$	50 mΩ	88 mΩ	60 mΩ
$I_{D,max}$	28 A	31 A	36 A
$V_{GS,max}$	10 V	30 V	19 V
$V_{th,min}$	1.1 V	3.5 V	1.8 V
$Q_{gs,on}$	0.7 nC	5 nC	7.5 nC
$Q_{gd,400V}$	2.2 nC	13 nC	14 nC

A GaN transistor half-bridge is utilized in this section as typical example to investigate the requirements for gate loop parasitics and the gate driver. Figure 2.10 shows the schematic of such a GaN half-bridge formed by a high-side (Q_{HS}) and a low-side (Q_{LS}) power transistor. Each of them is driven by its own gate driver. This driver can either be integrated on PCB or in the package together with the power transistor. Typically, a ceramic bypass capacitor C_{bp} with low equivalent series inductance and resistance is placed close to the driver providing high transient currents required for rapid gate driving. However, PCB traces and bondwires form a distributed, parasitic inductance L_p between C_{bp} and the driver as well as between the driver and the power transistor.

Gate Loop Requirements for Safe Turn-On

Figure 2.10a shows the current path $i_{turn-on}$ for charging the gate capacitance C_{gs} in order to turn on the low-side transistor. Along the current path, the series connection of C_{bp}, L_p and the gate-source capacitance C_{gs} of the power transistor forms a resonant tank. When the gate driver turns the power transistor on, this resonant tank is excited with a voltage step. The resulting ringing of the gate-source voltage V_{GS} may harm the sensitive gate of a GaN transistor [69, 70], which typically allows a maximum V_{GS} of 10 V for transient voltages [72] (see Table 2.2). This value is much lower than the maximum allowed transient V_{GS} of comparable silicon power transistors with up to 30 V [84] or 19 V for SiC transistors, respectively [120] (Table 2.2).

In order to protect the gate of GaN transistors, many implementations require an additional series resistor R_G between gate driver output and the gate of the GaN power transistor. This resistor acts as damping element in the resonant tank to avoid a gate voltage overshoot. However, it slows down the switching speed of the GaN transistor and may increase transition losses in a power converter system.

Fig. 2.10 Schematic of a gate
drive circuit for a discrete GaN
half-bridge for **a** low-side
HEMT turn-on and **b** slewing
of the switching node

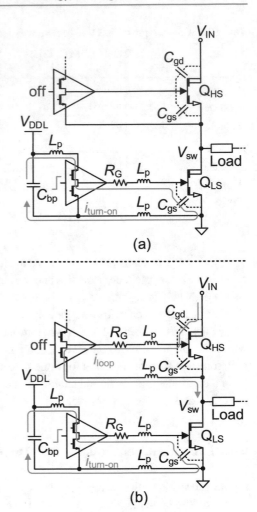

(a)

(b)

The behavior of the RLC resonant tank is defined by the differential equation for the
current which can be obtained by applying Kirchhoff's voltage law to the gate loop. It is
given in its general form by Eq. 2.8.

$$L\frac{di}{dt} + R \cdot i + \frac{1}{C}\int_0^t i\,dt = 0 \quad \rightarrow \quad s^2 + \frac{R}{L} \cdot s + \frac{1}{L \cdot C} = 0 \tag{2.8}$$

Applying the Laplace transformation, the differential equation can be simplified to a
quadratic equation for the Laplace variable s. The relation can be solved for s as given in
Eq. 2.9.

$$s = -\frac{R}{2L} \pm \sqrt{\left(\frac{R}{2L}\right)^2 - \frac{1}{LC}} \tag{2.9}$$

Different cases can be distinguished depending on nature of the radicand. For $\left(\frac{R}{2L}\right)^2 <$ $\frac{1}{LC}$, the radicand is negative leading to a complex value for s which indicates an oscillating behavior. The resonant tank is under-damped. For $\left(\frac{R}{2L}\right)^2 > \frac{1}{LC}$, the radicand is positive and the circuit shows a slow, over-damped behavior. If $\left(\frac{R}{2L}\right)^2 = \frac{1}{LC}$, the radicand is zero and the root expression vanishes. This leads to two identical solutions for s and is referred to as critically damped.

Based on the case distinction over-damped, under-damped, and critically damped, a damping factor D can be defined according to Equation 2.10. $D - 1$ corresponds to critical damping, $D < 1$ to an under-damped and $D > 1$ to an over-damped behavior.

$$D = \frac{R}{2} \cdot \sqrt{\frac{C}{L}} \equiv 1 \quad \rightarrow \quad R_\text{G} = 2 \cdot \sqrt{\frac{L_\text{p}}{C_\text{loop}}} \sim 2 \cdot \sqrt{\frac{L_\text{p}}{C_\text{gs}}} \qquad (2.10)$$

A transient simulation of the resonant tank is performed to estimate the destructive potential of gate voltage overshoots caused by resonant ringing at the GaN transistor gate. According to the parameters given in Table 2.2, C_gs is set to 242 pF. An optimized layout with 2.5 mm distance between C_bp and the driver and another 2.5 mm between driver and power transistor is assumed to estimate the value of the distributed gate loop inductance L_p. The values for these distances are estimated as a combination of PCB traces, pins, and bond-wires. For the full turn-on current loop marked in Fig. 2.10 (a), this sums up to a wiring length of 10 mm. Thus, the gate loop inductance L_p is around 10 nH if the general first-order approximation for parasitic inductances of PCB traces with 1 nH/mm is used. C_bp is set to high value of 2.2 μF. Hence, the voltage $V_\text{DDL} = 7\,$V is nearly constant during switching of the power transistor.

Figure 2.11 shows simulated waveforms or the gate-source voltage V_GS at turn-on for different values of R_G. For $R_\text{G} = 2.5\,\Omega$, the resonant tank is under-damped. The resulting ringing of V_GS leads to peak value of approximately 11 V which might destroy the power transistor since it exceeds the absolute maximum rating of 10 V (see Table 2.2).

For the given values, Eq. 2.10 results in $R_\text{G} = 12.8\,\Omega$ if critical damping $D = 1$ is desired. This is the smallest value for R_G to avoid any gate overshoot [121]. For larger values of R_G, the resonant tank is over-damped. In this case, any ringing and overshoot of V_GS is avoided, but the driver circuit has a slow switching characteristic (see "$R_\text{G} = 50\,\Omega$" in Fig. 2.11). This contradicts the desired high-speed switching behavior inherent to GaN transistors. For series gate resistances below 12.8 Ω, the gate voltage shows some ringing which may lead to a destructive voltage overshoot. For $R_\text{G} = 2.5\,\Omega$, the gate voltage rises faster than for $R_\text{G} = 12.8\,\Omega$ and reaches the nominal turn-on voltage of 7 V much faster. However, initially V_GS rises above 10 V (Fig. 2.11) and thereby exceeds the maximum allowed transient gate voltage (Table 2.2). This might cause destruction of the device.

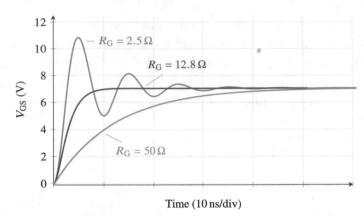

Fig. 2.11 Simulated V_{GS} for different values of the damping resistor R_G of the resonant tank

In practical approaches, some overshoot of the gate voltage can be tolerated by the GaN transistor. Therefore a damping resistor in the order of some 4 to 10Ω can be selected. This value is a bit lower than the resistance for critical damping calculated from Eq. 2.10. Thereby, higher switching speed is achieved while the overshoot at the transistor gate stays within the specified range.

Gate Loop Requirements for Safe Off-State

Besides voltage overshoots at the turn-on of GaN transistors, the parasitic gate loop inductance may also have negative effects when a transistor is turned off and the drain-source voltage V_{DS} changes. Figure 2.10b illustrates this case when Q_{HS} has just been turned off and the low-side transistor Q_{LS} starts to turn on. At that moment, V_{DS} of the high-side transistor experiences a fast transition from zero up to 400 V typically within a few nanoseconds. Similar to the low side, the same parasitic gate loop inductance L_p and therefore also the same R_G is present at the high side. During slewing of V_{DS}, it is imperative to keep the high-side transistor turned off in order to avoid destructive cross currents in the half-bridge when both transistors are turned on at the same time. For GaN transistors, this is especially critical, since the threshold voltage for turn-on can be as low as 1.1 V while comparable silicon superjunction transistors have threshold voltages above 3.5 V. Table 2.2 confirms a very low turn-on gate charge of 0.7 nC for a GaN transistor. At the same time, the gate-drain charge induced by the V_{DS} slewing from zero to 400 V is as large as 2.2 nC. Hence, Q_{HS} can be turned on by its own V_{DS} slewing when no additional measures are taken. Nevertheless, there are several options to prevent an unwanted turn-on of Q_{HS}. One straightforward solution is to artificially enlarge C_{gs} of the GaN transistor in order to increase the required turn-on gate charge $Q_{gs,on}$ to a value greater than the Miller charge coupled into the gate through c_{gd}. The required C_{gs} for that purpose can be calculated according to Equation 2.11.

$$Q_{gs,on} = C_{gs} \cdot V_{th} > Q_{gd,400V}$$

$$C_{GS} > \frac{Q_{gd,400V} - Q_{gs,on}}{V_{th}} = 1.36\text{nF} \qquad (2.11)$$

The capacitor providing the required additional C_{gs} has to be placed as close as possible to the GaN transistor to avoid transient effects by parasitic inductances. As intended, the additional capacitor increases the required gate charge for turn-on improving the robustness of the off-state. On the down side, it slows down the turn-on speed of the GaN transistor contradicting the use of GaN for its high-speed switching performance. Additionally, this also increases the switching losses related to the gate charge. If the additional gate capacitance cannot be tolerated, the maximum values for the gate loop impedance have tight limits to avoid unwanted turn-on of the power transistor. When no large gate capacitor is available to take the gate-drain charge $Q_{gd,400V}$, it has to be dissipated through the gate loop (i_{loop} in Fig. 2.10b). Figure 2.12 illustrates an equivalent circuit model for a GaN transistor during slewing of V_{DS}. It incorporates C_{gs}, the parasitic gate loop inductance L_p and the total gate series resistance composed of R_G and the pull-down resistance of the gate driver R_{drv}. During slewing of V_{DS}, a capacitive current is generated by the Miller capacitance c_{gd}. It is modeled by the current source I_{dg}. As a first-order approximation it is assumed that this current is constant during slewing. Hence, it is defined as the total Miller charge divided by the duration of the transition $I_{dg} = Q_{gd,400V}/tf$.

In the following, absolute maximum values for the inductance as well as for the resistance in the gate loop are examined separately. First, the inductance L_p is assumed to be zero in order to obtain an absolute maximum value for the gate loop resistance avoiding any unwanted turn-on of the power transistor. Therefore, voltage drop across the resistance in the gate loop caused by the capacitive current I_{dg} has to stay below the threshold voltage V_{th} of the transistor. This relation is given by Eq. 2.12.

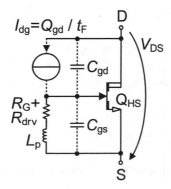

Fig. 2.12 Equivalent circuit model for the high-side transistor during slewing

$$R_G + R_{drv} < \frac{V_{th}}{\left(Q_{gd,400V} - Q_{GS,on}\right)/T_f} = \frac{1.1V}{0.29A} = 3.8\Omega \qquad (2.12)$$

For a fast transition time $tf = 5ns$ achievable by using GaN transistors, the maximum allowable value for $R_G + R_{drv}$ is 3.8 Ω. Such a low value is difficult to achieve, since R_G has to be greater than 4Ω to avoid destructive gate overshoot voltages during turn-on according to Eq. 2.10.

In a separate consideration, the absolute maximum value for L_p is calculated. Therefore, the resistive part of the gate loop impedance is neglected and $R_G + R_{drv}$ is assumed to be zero. When V_{DS} of the power transistor begins to rise, the resonant tank formed by C_{gs} and L_p is excited by a current step from zero to $I_{dg} = 0.29A$. The maximum value for L_p to ensure $V_{GS} < V_{th}$ can be calculated by solving the energy equation for a LC-resonant tank as given in Eq. 2.13.

$$\frac{1}{2}L \cdot I^2 = \frac{1}{2}C \cdot V^2 \quad \rightarrow \quad L_p < \frac{C_{gs} \cdot V_{th}^2}{I_{dg}^2} = 1.01nH \qquad (2.13)$$

Achieving a gate loop inductance below 1 nH is nearly impossible if a discrete gate driver is used, since already 1 mm PCB trace or bond wire shows such an inductance value.

It is very challenging to fulfill the conditions from Eqs. 2.12 and 2.13 without an additional gate capacitor based on Eq. 2.11. The minimum required R_G to avoid ringing (Eq. 2.10) is larger than the maximum allowed R_G to avoid turn on during switching transitions (Eq. 2.12). Additionally, it is very difficult to meet the maximum allowed L_p to avoid self turn on (Eq. 2.13).

In order to avoid additional gate capacitance with its negative influence on switching speed and efficiency, bipolar or three-level gate drivers are proposed for GaN transistors [114]. They can provide a negative gate-source voltage to increase the margin towards V_{th} and, thereby, keep the transistor safely off. Thus, V_{th} in Eqs. 2.11, 2.12, and 2.13 extends to $V_{th} - V_{GS,off}$ with $V_{GS,off} < 0V$. This buys some freedom for gate loop parasitics and allows higher values for R_G and L_p without additional gate capacitance. When the transistor is turned off using a typical voltage $V_{GS,off} = -5V$, the resistance $R_G + R_{drv}$ can be as large as 21Ω (from Eq. 2.12) and the maximum allowable L_p increases from 1nH to 31 nH (from Eq. 2.13).

A bipolar gate driver provides a negative V_{GS} during the whole off-time. For GaN transistors, this leads to high losses in hard switching operation with an inductive load. Since GaN transistors do not possess a body diode, they have to operate in third-quadrant reverse conduction with positive V_{GD} and negative V_{DS} during inductive switching until the transistor is turned on (see Sect. 2.3). When a negative V_{GS} is applied, the third-quadrant conduction losses increase considerably. Therefore, three-level gate drivers are proposed, which apply the negative gate-source voltage only during switching transitions and provide $V_{GS} = 0V$ for the steady off-state [114]. Thus, third-quadrant conduction losses are reduced [122].

While three-level gate driving is a good method for well-controlled switching and safe on- and off-states, it comes at the expense of considerable design overhead. They typically require a full-bridge configuration of the driver output stage to provide the different gate-source voltages required for the power transistor. An elegant alternative is to minimize the gate loop parasitics as root cause for the challenges of GaN driving. This can be achieved by reducing the distance between driver and power transistor close to zero by monolithic integration of the gate driver and the power transistor on one IC. Thereby, L_p is minimized to a value close to zero, directly fulfilling the condition in Eq. 2.13. Additionally, the value for R_G to achieve critical damping of the resonant tank (Eq. 2.10) reduces and thereby the condition for the maximum R_G as given in Eq. 2.12 can be met more easily. Thus, monolithic integration as presented in this work combines several benefits:

- Reduced number of dies.
- Reduced system complexity on PCB.
- Improved reliability due to lower part count.
- Simplified gate driver design due to minimized gate loop parasitics.

In order to exploit these benefits, increasing levels of integration around GaN power transistors are commercially available or published in academic and corporate research. Section 2.5 examines different integration levels on the path towards monolithically integrated power converter ICs.

2.5   Levels of Integration in GaN

The investigations in Sect. 2.4 suggest that the parasitic gate loop inductance can pose significant challenges to gate drivers for GaN transistors. In order to simplify the gate driver design and avoid the need for complex driver architectures such as bipolar and multi-level gate driving, the gate loop inductance has to be minimized. This section presents a review on different approaches and levels of integration that minimize the gate loop parasitics and maximize the utilization of the superior switching characteristics of GaN for compact and efficient power electronics.

Package Integration

One way to minimize the gate loop parasitics is to integrate a silicon driver IC together with a high-voltage GaN power transistor in a single package. Thereby, the distance between driver and power transistor is reduced compared to a discrete placement on PCB. Consequently, the gate loop inductance is decreased. The lack of pins or even leads further lowers the parasitic inductances. Thus, it should be possible to achieve a gate loop inductance below 1

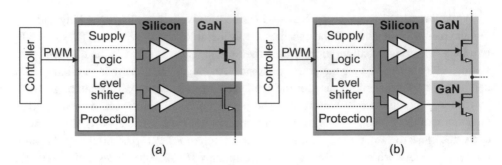

Fig. 2.13 Block diagram of commercially available package integration solutions with GaN transistors: **a** single d-mode transistor as HV cascode [61], **b** two e-mode transistors in a half-bridge configuration [125]

nH by package integration of driver and GaN transistor. This remaining parasitic inductance is related to the connections between driver and transistor inside of the package, which can be implemented as inter-die bonds or as a substrate included in the package for wiring purposes.

Figure 2.13 shows the block diagrams of two commercially available GaN solutions utilizing package integration. In [61] (Fig. 2.13a), a 600 V d-mode GaN HEMT is used as a high-voltage cascode for a low-voltage silicon e-mode transistor according to Fig. 2.2b. The gate driver is integrated in the same package on one silicon die together with the silicon power transistor [123]. The driver includes various features such as level shifting, protection circuits (i.e. under-voltage lockout (UVLO), and over-current protection (OCP)) as well as a supply generation concept. It is designed in a 180 nm silicon technology [124], which is more mature than GaN and where broad design experience exists. Thus, it can benefit from the maturity of silicon technology and may include proven circuits to achieve good accuracy such as a bandgap voltage reference and general trimming of critical parameters. However, the package integrates only the power stage including the power transistor and the gate driver. An additional controller is required to implement a power converter.

Similar to that, Fig. 2.13b shows another example of a package-integrated power stage [125]. Two e-mode GaN transistors are enclosed in one package together with a high-voltage half-bridge driver [126]. This driver IC also contains various features including a bootstrap rectifier to generate a floating supply voltage for the high-side driver and several protection circuits. However, integrating multiple dies in one package, especially with isolation requirements of several hundred volts between the dies, leads to significant package complexity [126]. This increases the effort for assembly and test as well as the associated cost. Nevertheless, due to the limited risk by designing all analog, digital, and mixed-signal circuits in mature silicon, package integration seems to be the approach chosen by established semiconductor companies in the power management field.

Power Stage Integration in GaN

The lateral structure of GaN power transistors allows for monolithic integration of multiple components on one die (see Sect. 2.1). Thus, integrating the gate driver in GaN technology on one die together with the high-voltage power transistor is possible and nearly eliminates gate loop parasitics. Due to the limited maturity level of GaN technology and especially due to the lack of suitable complementary p-type devices (see Sect. 2.2), the design of a gate driver in GaN is challenging. Nevertheless, some solutions with a GaN-integrated power stage are commercially available.

Figure 2.14 shows the block diagrams of two power stages, which are integrated monolithically in an e-mode GaN technology. Figure 2.14a shows a monolithic integration of a single 650 V power transistor together with the gate driver and a supply regulator to generate the bias voltages for the gate driver [127]. This supply regulator, however, requires an auxiliary Zener diode on PCB as voltage reference. Additionally, no protection features such as UVLO, OCP, or over-temperature protection (OTP) are included and an external controller is required to implement a power converter. Improvement of GaN-integrated gate drivers is an ongoing research topic and also a part of this work (see Sect. 4.2). Reference [75] presents different gate driver approaches aiming to increase the switching speed and the V_{DS} slew rate. Reference [128] proposes some enhancements for gate driver circuits in GaN including a temperature compensation in order to mitigate the influences of the considerably high temperature coefficient of GaN transistors (see Sect. 2.1). Additionally, this gate driver includes a voltage reference for the integrated supply regulator. Thus, the need for an external Zener diode including the required pads can be eliminated reducing the package complexity.

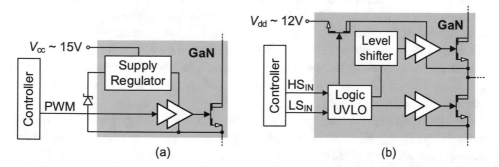

(a)   (b)

Fig. 2.14 Block diagram of commercially available GaN-integrated solutions: **a** single e-mode transistor with driver and supply regulator [127], **b** two e-mode transistors in a half-bridge configuration with driver and bootstrap rectifier [129]

Figure 2.14b depicts a monolithic half-bridge including both power transistors together with the drivers and the level shifter for the high-side transistor on one GaN die [129].

Table 2.3 Comparison of examples for different levels of GaN integration

Affiliation	[61] industry	[125] industry	[127] industry	[129] industry	[75] academia	[128] academia
Technology	d-GaN + Si	e-GaN	e-GaN	e-GaN	e-GaN	e-GaN
GaN integration	HEMT	HEMT	Power stage[a]	Power stage[b]	Power stage[a]	Power stage[a]
Topology	Single	Half-bridge	Single	Half-bridge	Single	Single
# dies in package	2	3	1	1	1	1
Year	2018	2020	2020	2020	2018	2021
BVDSS (V)	600	650	650	70	650	650
$R_{DS,On}$ (mΩ)	57	150	70	8.5	130	n/a
max. I_D (A)	34	10	20	12.5	>20	n/a
max. f_{sw} (mH)	0.5	2	2	3	n/a	50
max. dV_{DS}/dt (V/ns)	>150	100	200	<100	335	118

[a] HEMT and gate driver, [b] HEMT, gate driver and level shifter

Additionally, a bootstrap rectifier for the generation of a floating supply voltage required to drive the high-side transistor is included. Additionally, basic protection is implemented by a UVLO as well as some logic to ensure non-overlap operation of the transistors forming a half-bridge. This half-bridge, however, is limited to 60 V input voltage and therefore not applicable for operation directly from the AC power grid. It also requires an external controller on PCB to implement a power converter. In 2017, a press release announced a package-integrated half-bridge with two GaN dies rated for 650 V [130]. However, no official data sheet is available, yet. This can be seen as a hint for the challenges towards integrating a 650 V half-bridge using only GaN technology, which are not yet reliably solved (see also Sect. 2.3).

The existing solutions with integrated power stages in GaN technology prove the possibility and viability of integration in GaN for efficient and compact power electronics. Nevertheless, this approach is so far pursued only by fab-less startup companies and not yet by established semiconductor companies. Table 2.3 provides a summarizing comparison of the different integration approaches around GaN technology. References [61, 125, 127, 129] is even commercially available and, thus, developed until productization.

Monolithic GaN Integration

In order to reduce PCB complexity and achieve compact and easy-to-use power converters, an even higher level of integration with integrated control loop is desired. Especially for the high-speed operation of GaN transistors, this allows for short signal paths and fast response of the control thanks to sensing capabilities inside the power stage. While an integrated control loop limits the flexibility of the implementation, it makes the solution very easy to use. In silicon technology, many companies offer power converter ICs integrating high-voltage power transistors, gate drivers, sensing circuits, and control loops [131–133]. Typically, various protection features as well as a high-voltage startup and a supply voltage generation are included to minimize the required external components and the PCB complexity.

In e-mode GaN technology, monolithic integration of a full power converter IC is mainly an academic research topic, yet. Due to the various process-related challenges (see Sect. 2.2), research on circuit design in GaN is required to enable fully integrated power converter ICs in GaN. One set of challenges is related to the lack of suitable p-type devices, which limits the achievable gain of amplifier stages, increases voltage head-room challenges, and causes higher quiescent power consumption (see also Sect. 3.1 and Sect. 3.2). Another set of challenges is related to the defect density of the GaN epitaxial layer (Sect. 2.2), which limits intrinsic device matching (see also Sect. 3.1). Last but not least, the lack of well-controlled pn junctions prevents the established approach of a bandgap circuit as voltage and current references. Hence, new design techniques have to be investigated and developed to make integration in GaN a viable option. Several publications present single or multiple circuit blocks required for integrated power converter ICs such as

- voltage reference circuits [128, 134, 135];
- sensing circuits for current [136, 137] and temperature [138];
- saw tooth signal generators [139] and full pulse width modulation (PWM) generators [66] based on simple comparator circuits;
- protection circuits to prevent gate voltage overshoots [140] and over-current [141] at the power transistor; and
- various digital logic gates using resistors and d-mode pull-up [142] as well as a ring oscillator based on complementary devices [90].

Various design considerations for several GaN-integrated circuits overcoming the challenges posed by the process technology are provided in Chap. 3. Where applicable, the circuits presented in the itemization above are presented and discussed in more detail.

Few conference publications claim all-GaN ICs for power conversion. Reference [143] presents a GaN power IC platform and an all-GaN DC-DC buck converter IC. However, for operation at least five different reference voltages have to be provided to the IC and, additionally, an external supply voltage has to be applied for the integrated circuits. Furthermore, the presented circuit is only able to handle comparably low input voltages smaller than 30 V and power efficiency values are not provided. No related article has been published by the authors in an IEEE peer-reviewed journal. Another research group presented an all-GaN IC with a monolithically integrated half-bridge enabled by substrate isolation [79]. This implementation requires only two external voltage references in addition to a supply voltage. The publication includes measured transient waveforms for the power stage with 1 MHz switching operation at 200 V input. However, measurements are only published for the power stage itself. For the full converter IC with the integrated control loop, only simulated waveforms are provided and no measurements are available.

The highest level of integration with measured transient waveforms proving operation of a monolithically integrated GaN IC for power converters is presented in publications related to this work [144]. This IC integrates a 650 V GaN transistor together with its gate driver, an analog control loop, a supply voltage regulator, and basic protection features such as UVLO, OCP, and an electrostatic discharge (ESD) active clamp. It shows transient waveforms of clocking operation at 260 kHz with 400 V input voltage and achieves an efficiency up to 95.6%. The conference paper was featured in a press article [145] and the authors also published an extended IEEE journal paper [80]. Detailed design considerations on circuit and system level as well as characterization results for both, core design blocks and the full converter, are presented in Chap. 4. A detailed comparison with a silicon power converter IC is provided in Chap. 5.

References

1. Asif Khan, M., et al. (1993). High electron mobility transistor based on a GaNAlxGa1- xN heterojunction. *Applied Physics Letters, 63*(9), 1214–1215. https://doi.org/10.1063/1.109775
2. Asif Khan, M., et al. (1996). Enhancement and depletion mode GaN/AlGaN heterostructure field effect transistors. *Applied Physics Letters, 68*(4), 514–516. https://doi.org/10.1063/1.116384
3. Sheppard, S. T., et al. (2002). High power hybrid and MMIC amplifiers using wide-bandgap semiconductor devices on semi-insulating SiC substrates. In *60th DRC. Conference Digest Device Research Conference* (pp. 175–178).
4. Kikkawa, T., et al. (2004). An over 200-W output power GaN HEMT push-pull amplifier with high reliability. In *2004 IEEE MTT-S International Microwave Symposium Digest* (IEEE Cat. No.04CH37535) (Vol. 3, pp. 1347–1350).
5. Yamanaka, K., et al. (2005). S and C Band over 100 W GaN HEMT 1-chip high power amplifiers with cell division configuration. *European Gallium Arsenide and Other Semiconductor Application Symposium, GAAS, 2005*, 241–244.
6. Li, S., et al. (2018a). Design of a Compact GaN MMIC doherty power amplifier and system level analysis with X-parameters for 5G communications. *IEEE Transactions on Microwave Theory and Techniques, 66*(12), 5676–5684.
7. Wong, J., Watanabe, N., & Grebennikov, A. (2018). High-power high-efficiency broadband GaN HEMT Doherty amplifiers for base station applications. In *2018 IEEE Topical Conference on RF/Microwave Power Amplifiers for Radio and Wireless Applications (PAWR)* (pp. 16–19).
8. Raza, A., & Gengler, J. (2019). Design of a 110 W wideband inverse class-F GaN HEMT power amplifier with 65% efficiency over 100-1000 MHz bandwidth. In *2019 IEEE Topical*

Conference on RF/Microwave Power Amplifiers for Radio and Wireless Applications (PAWR) (pp. 1–4).

9. Avrutin, V., et al. (2010). Growth of bulk GaN and AlN: Progress and challenges. *Proceedings of the IEEE, 98*(7), 1302–1315.

10. SystemPlus Consulting. (2019). GaN-on-sapphire HEMT power IC by Power Integrations - InnoSwitch3 Flyback Switcher Power IC in Anker PowerPort A2017 (Sample). https://www.systemplus.fr/wp-content/uploads/2019/07/SP19480-Power-Integrations-GaN-on-Sapphire-HEMT_sample.pdf. Retrieved from 04/06/2021.

11. Chen, J.-T., et al. (2018). A GaN-SiC hybrid material for high-frequency and power electronics. *Applied Physics Letters, 113*(4), 041605. https://doi.org/10.1063/1.5042049

12. Chumbes, E. M., et al. (1999). Microwave performance of AlGaN/GaN high electron mobility transistors on Si (111) substrates. In *International Electron Devices Meeting 1999*. Technical Digest (Cat. No.99CH36318) (pp. 397–400).

13. Saito, W., et al. (2007). Suppression of dynamic on-resistance increase and gate charge measurements in high-voltage GaN-HEMTs with optimized field-plate structure. *IEEE Transactions on Electron Devices, 54*(8), 1825–1830. https://doi.org/10.1109/TED.2007.901150

14. Brown, D. F., et al. (2013). High-speed, enhancement-mode GaN power switch with regrown n+ GaN ohmic contacts and staircase field plates. *IEEE Electron Device Letters, 34*(9), 1118–1120. https://doi.org/10.1109/LED.2013.2273172

15. Chiu, H., et al. (2013). Characteristics of AlGaN/GaN HEMTs with various field-plate and gate-to-drain extensions. *IEEE Transactions on Electron Devices, 60*(11), 3877–3882. https://doi.org/10.1109/TED.2013.2281911

16. Kaneko, S., et al. (2015). Current-collapse-free operations up to 850 V by GaNGIT utilizing hole injection from drain. In: *2015 IEEE 27th International Symposium on Power Semiconductor Devices IC's (ISPSD)* (pp. 41–44). https://doi.org/10.1109/ISPSD.2015.7123384

17. Mohammad, S. N., Salvador, A. A., & Morkoc, H. (1995). Emerging gallium nitride based devices. *Proceedings of the IEEE, 83*(10), 1306–1355.

18. Ambacher, O. (1998). Growth and applications of group III-nitrides. *Journal of Physics D: Applied Physics, 31*(20), 2653–2710. https://doi.org/10.1088/0022-3727/31/20/001, https://doi.org/10.1088

19. Kaminski, N. (2009). State of the art and the future of wide band-gap devices. In *2009 13th European Conference on Power Electronics and Applications* (pp. 1–9).

20. Elasser, A., & Chow, T. P. (2002). Silicon carbide benefits and advantages for power electronics circuits and systems. *Proceedings of the IEEE, 90*(6), 969–986.

21. Wang, Y., et al. (2018). High-uniformity and high drain current density enhancement-mode AlGaN/GaN gates-seperating groove HFET. *IEEE Journal of the Electron Devices Society, 6*, 106–109.

22. Barker, J. M., et al. (2005). Bulk GaN and AlGaN/GaN heterostructure drift velocity measurements and comparison to theoretical models. *Journal of Applied Physics, 97*(6), 063701–5. https://doi.org/10.1063/1.1854724

23. Baliga, B. J. (1982). semiconductors for high-voltage, vertical channel field- effect transistors. *Journal of Applied Physics, 53*(3), 1759–1764. https://doi.org/10.1063/1.331646.

24. Baliga, B. J. (1996). Trends in power semiconductor devices. *IEEE Transactions on Electron Devices, 43*(10), 1717–1731.

25. Baliga, B. J. (1989). Power semiconductor device figure of merit for high-frequency applications. *IEEE Electron Device Letters, 10*(10), 455–457.

26. Bahat-Treidel, E., et al. (2010). AlGaN/GaN/GaN: C back-barrier HFETs with breakdown voltage of over 1 kV and low $R_{ON} \times A$. *IEEE Transactions on Electron Devices, 57*(11), 3050–3058.

27. Shibata, D., et al. (2016). 1.7 kV/1.0mΩcm2 normally-off vertical GaN transistor on GaN substrate with regrown p-GaN/AlGaN/GaN semipolar gate structure. In *2016 IEEE International Electron Devices Meeting (IEDM)* (pp. 10.1.1–10.1.4).

28. Lee, J., et al. (2018). High figure-of-merit (V_{BR}^2/R_{ON}) AlGaN/GaN power HEMT with periodically C-doped GaN buffer and AlGaN back barrier. *IEEE Journal of the Electron Devices Society, 6*, 1179–1186.

29. Lidow, A. et al. (2012). *GaN transistors for efficient power conversion*, 1st. edn. El Segundo, CA: Power Conversion Publications. 208 pp. ISBN: 9780615569253.

30. Kuzuhara, M., & Tokuda, H. (2015). Low-loss and high-voltage III-nitride transistors for power switching applications. *IEEE Transactions on Electron Devices, 62*(2), 405–413.

31. Meneghini, M., Meneghesso, G., & Zanoni, E. (Eds.) (2017). *Power GaN Devices*. Switzerland: Springer International Publishing. 380 pp. ISBN: 9783319431970. https://doi.org/10.1007/978-3-319-43199-4.

32. Visalli, D., et al. (2009). AlGaN/GaN/AlGaN double heterostructures on silicon substrates for high breakdown voltage field-effect transistors with low on-resistance. *Japanese Journal of Applied Physics, 48*(4), 04C101. https://doi.org/10.1143/jjap.48.04c101, https://doi.org/10.1143

33. Ota, K., et al. (2009). A normally-off GaN FET with high threshold voltage uniformity using a novel piezo neutralization technique. In *2009 IEEE International Electron Devices Meeting (IEDM)* (pp. 1–4).

34. Morita, T., et al. (2007). 650 V 3.1 mΩcm2 GaN-based monolithic bidirectional switch using normally-off gate injection transistor. In *2007 IEEE International Electron Devices Meeting* (pp. 865–868).

35. Boutros, K. S., et al. (2009). Normally-off 5A/1100V GaN-on-silicon device for high voltage applications. In *2009 IEEE International Electron Devices Meeting (IEDM)* (pp. 1–3).

36. Medjdoub, F., et al. (2010). Low on-resistance high-breakdown normally off AlN/GaN/AlGaN DHFET on Si substrate. *IEEE Electron Device Letters, 31*(2), 111–113.

37. Lu, B., & Palacios, T. (2010). High breakdown (> 1500 V) AlGaN/GaN HEMTs by substrate-transfer technology. *IEEE Electron Device Letters, 31*(9), 951–953.

38. Morita, T., et al. (2011). 99.3% efficiency of three-phase inverter for motor drive using GaN-based gate injection transistors. In *2011 Twenty-Sixth Annual IEEE Applied Power Electronics Conference and Exposition (APEC)* (pp. 481–484).

39. Sun, M., et al. (2012). Comparative breakdown study of mesa- and ion-implantation-isolated AlGaN/GaN high-electron-mobility transistors on Si substrate. *Applied Physics Express, 5*(7), 074202. https://doi.org/10.1143/apex.5.074202.

40. Jiang, Q., et al. (2013). 1.4-kVAlGaN/GaN HEMTs on a GaN-on-SOI platform. *IEEE Electron Device Letters, 34*(3), 357–359.

41. Posthuma, N. E., et al. (2018). An industry-ready 200 mm p-GaN E-mode GaNon- Si power technology. In *2018 IEEE 30th International Symposium on Power Semiconductor Devices and ICs (ISPSD)* (pp. 284–287).

42. Germain, M., et al. (2010). GaN-on-Si power field effect transistors. In *Proceedings of 2010 International Symposium on VLSI Technology, System and Application* (pp. 171–172).

43. Li, X., et al. (2017). 200 V enhancement-mode p-GaN HEMTs fabricated on 200 mm GaN-on-SOI with trench isolation for monolithic integration. *IEEE Electron Device Letters, 38*(7), 918–921.

44. Uemoto, Y., et al. (2006). A normally-off AlGaN/GaN transistor with RonA = 2.6mΩcm2 and BVds=640V using conductivity modulation. In *2006 International Electron Devices Meeting* (pp. 1–4).

45. Dora, Y., et al. (2006). High breakdown voltage achieved on AlGaN/GaN HEMTs with integrated slant field plates. *IEEE Electron Device Letters, 27*(9), 713–715.
46. Handa, H., et al. (2016). High-speed switching and current-collapse-free operation by GaN gate injection transistors with thick GaN buffer on bulk GaN substrates. In: *2016 IEEE International Electron Devices Meeting (IEDM)* (pp. 10.3.1–10.3.4).
47. Gupta, C., et al. (2017). 1 kV field plated in-situ oxide, GaN interlayer based vertical trench MOSFET (OG-FET). In *2017 75th Annual Device Research Conference (DRC)* (pp. 1–2).
48. HoKwan, M., et al. (2014). CMOS-compatible GaN-on-Si field-effect transistors for high voltage power applications. In *2014 IEEE International Electron Devices Meeting* (pp. 17.6.1–17.6.4).
49. Hwang, I., et al. (2012). 1.6kV, 2.9 mΩcm2 normally-off p-GaN HEMT device. In *2012 24th International Symposium on Power Semiconductor Devices and ICs* (pp. 41–44).
50. Hilt, O., et al. (2010). Normally-off AlGaN/GaN HFET with p-type Ga Gate and AlGaN buffer. In *2010 22nd International Symposium on Power Semiconductor Devices IC's (ISPSD)* (pp. 347–350).
51. Saito, W., et al. (2006). Recessed-gate structure approach toward normally off high-voltage AlGaN/GaN HEMT for power electronics applications. *IEEE Transactions on Electron Devices, 53*(2), 356–362.
52. Chu, R., et al. (2011). 1200-V normally Off GaN-on-Si field-effect transistors with low dynamic on-resistance. *IEEE Electron Device Letters, 32*(5), 632–634.
53. Nie, H., et al. (2014). 1.5-kV and 2.2mΩ–cm^2 vertical GaN transistors on bulk-GaN substrates. *IEEE Electron Device Letters, 35*(9), 939–941.
54. Ji, D., et al. (2018). 880 V 2.7mΩ · cm^2 MIS gate trench CAVET on bulk GaN substrates. *IEEE Electron Device Letters, 39*(6), 863–865.
55. Neufeld, C., & Wu, Y. (2020). Pushing the boundaries of high voltage GaN power conversion. https://www.eeworldonline.com/pushing-theboundaries-of-high-voltage-gan-power-conversion/. Retrieved from 06 April 2021.
56. Infineon. (2011). Rev. 2.0, 650V CoolMOS™CFD2 power transistor. In *IPx65R150CFD Datasheet*. https://www.infineon.com/dgdl/Infineon-IPX65R150CFD-DS-v02_00-en.pdf?fileId=db3a3043338c8ac80133ace218433063.
57. Infineon. (2014). Rev. 2.0, 600V CoolMOS™P6 power transistor. In *IPW60R070P6 Datasheet*. https://www.infineon.com/dgdl/Infineon-IPW60R070P6-DS-v02_00-en.pdf?fileId=5546d461464245d3014694ab3f43692e.
58. Infineon. (2020). Rev. 2.11, 600V CoolGaN™enhancement-mode power transistor. In *IGO60R070D1 Datasheet*. https://www.infineon.com/dgdl/Infineon-IGO60R070D1-DataSheet-v02_11-EN.pdf?fileId=5546d46265f064ff016685f053216514.
59. GaN Systems. (Rev. 200423, 2009–2020). Bottom-side cooled 650V E-mode GaN transistor. In *GS-065-011-1-L Preliminary Datasheet*. https://gansystems.com/wp-content/uploads/2020/04/GS-065-011-1-L-DSRev-200423.pdf.
60. Ambacher, O., et al. (2000). Two dimensional electron gases induced by spontaneous and piezoelectric polarization in undoped and doped AlGaN/GaN heterostructures. *Journal of Applied Physics, 87*(1), 334–344. https://doi.org/10.1063/1.371866.
61. Texas Instruments. (2018). rev. 2020. LMG341xR050 600-V 50-m? integrated GaN FET power stage with overcurrent protection. In *LMG3410R050, LMG3411R050 Datasheet*. http://www.ti.com/lit/ds/symlink/lmg3410r050.pdf?ts=1588080212431.
62. Cai, Y., et al. (2005). High-performance enhancement-mode AlGaN/GaN HEMTs using fluoride-based plasma treatment. *IEEE Electron Device Letters, 26*(7), 435–437.

63. Wang, M., & Chen, K. J. (2011). Improvement of the off-state breakdown voltage with fluorine ion implantation in AlGaN/GaN HEMTs. *IEEE Transactions on Electron Devices, 58*(2), 460–465.
64. Chen, K. J., et al. (2011). Physics of fluorine plasma ion implantation for GaN normally-Off HEMT technology. In *2011 International Electron Devices Meeting* (pp. 19.4.1–19.4.4).
65. Wong, K., Chen, W., & Chen, K. J. (2009). Integrated voltage reference and comparator circuits for GaN smart power chip technology. In *2009 21st International Symposium on Power Semiconductor Devices IC's* (pp. 57–60).
66. Wang, H., et al. (2015). A GaN pulse width modulation integrated circuit for GaN power converters. *IEEE Transactions on Electron Devices, 62*(4), 1143–1149.
67. Chu, R., et al. (2012). Normally-off GaN-on-Si metal-insulator-semiconductor field-effect transistor with 600-V blocking capability at 200°C. In *2012 24th International Symposium on Power Semiconductor Devices and ICs* (pp. 237–240).
68. Hwang, I., et al. (2013). p-GaN gate HEMTs with tungsten gate metal for high threshold voltage and low gate current. *IEEE Electron Device Letters, 34*(2), 202–204.
69. Wu, T., et al. (2015). Forward bias gate breakdown mechanism in enhancement- mode p-GaN gate AlGaN/GaN high-electron mobility transistors. *IEEE Electron Device Letters, 36*(10), 1001–1003. https://doi.org/10.1109/LED.2015.2465137.
70. Rossetto, I., et al. (2016). Experimental demonstration of weibull distributed failure in p-type GaN high electron mobility transistors under high forward bias stress. In *2016 28th International Symposium on Power Semiconductor Devices and ICs (ISPSD)* (pp. 35–38). https://doi.org/10.1109/ISPSD.2016.7520771.
71. EPC. (2021). EPC2019 - enhancement mode power transistor. In *EPC2019 Datasheet*. https://epc-co.com/epc/Portals/0/epc/documents/datasheets/EPC2019_datasheet.pdf.
72. GaN Systems. (Rev. 200402, 2009–2020). Bottom-side cooled 650V E-mode GaN transistor. In *GS66508B Datasheet*. https://gansystems.com/wp-content/uploads/2020/04/GS66508B-DS-Rev-200402.pdf.
73. Udrea, F., Deboy, G., & Fujihira, T. (2017). Superjunction power devices, history, development, and future prospects. *IEEE Transactions on Electron Devices, 64*(3), 713–727.
74. Agarwal, A., Han, K., & Baliga, B. J. (2019). 600 V 4H-SiC MOSFETs fabricated in commercial foundry with reduced gate oxide thickness of 27 nm to achieve IGBT-compatible gate drive of 15 V. *IEEE Electron Device Letters, 40*(11), 1792–1795.
75. Tang, G., et al. (2018). High-speed, high-reliability GaN power device with integrated gate driver. In *2018 IEEE 30th International Symposium on Power Semiconductor Devices and ICs (ISPSD)* (pp. 76–79).
76. Xue, L., & Zhang, J. (2017). Active clamp flyback using GaN power IC for power adapter applications. In *2017 IEEE Applied Power Electronics Conference and Exposition (APEC)* (pp. 2441–2448).
77. Xue, L., & Zhang, J. (2018). Design considerations of highly-efficient active clamp flyback converter using GaN power ICs. In *2018 IEEE Applied Power Electronics Conference and Exposition (APEC)* (pp. 777–782).
78. Chen, K. J., et al. (2017). GaN-on-Si power technology: Devices and applications. *IEEE Transactions on Electron Devices, 64*(3), 779–795.
79. Li, X., et al. (2019). GaN-on-SOI: Monolithically integrated all-GaN ICs for power conversion. In *2019 IEEE International Electron Devices Meeting (IEDM)* (pp. 4.4.1–4.4.4). https://doi.org/10.1109/IEDM19573.2019.8993572.
80. Kaufmann, M., & Wicht, B. (2020). A monolithic GaN-IC with integrated control loop achieving 95.6%

81. Menozzi, R., et al. (2008). Temperature-dependent characterization of Al- GaN/GaN HEMTs: Thermal and source/drain resistances. *IEEE Transactions on Device and Materials Reliability, 8*(2), 255–264. https://doi.org/10.1109/TDMR.2008.918960.
82. Aminbeidokhti, A., et al. (2016). The power law of phonon-limited electron mobility in the 2-D electron gas of AlGaN/GaN heterostructure. *IEEE Transactions on Electron Devices, 63*(5), 2214–2218. https://doi.org/10.1109/TED.2016.2544920.
83. Batista, J., Mandelis, A., & Shaughnessy, D. (2003). Temperature dependence of carrier mobility in Si wafers measured by infrared photocarrier radiometry. *Applied Physics Letters, 82*(23), 4077–4079. https://doi.org/10.1063/1.1582376.
84. 84. Infineon. (Rev. 2.0, 2020). OSFET 600V CoolMOS™CFD7 power transistor. In *IPDD60R105CFD7 Datasheet*. https://www.infineon.com/dgdl/Infineon-IPDD60R105CFD7-DataSheet-v02_00-EN.pdf?fileId=5546d4627506bb3201751c00b00c424e.
85. Hahn, H., et al. (2014). First monolithic integration of GaN-based enhancement mode n-channel and p-channel heterostructure field effect transistors. In *72nd Device Research Conference* (pp. 259–260). https://doi.org/10.1109/DRC.2014.6872396.
86. Chowdhury, N., et al. (2019). p-channel GaN transistor based on p-GaN/AlGaN/GaN on Si. *IEEE Electron Device Letters, 40*(7), 1036–1039.
87. Zheng, Z., et al. (2020). High I_{ON} and I_{ON} / I_{OFF} ratio enhancement-mode buried p -channel GaN MOSFETs on p -GaN gate power HEMT platform. *IEEE Electron Device Letters, 41*(1), 26–29.
88. Chu, R., et al. (2016). An experimental demonstration of GaN CMOS technology. *IEEE Electron Device Letters, 37*(3), 269–271. https://doi.org/10.1109/LED.2016.2515103.
89. Chowdhury, N., et al. (2020). Regrowth-free GaN-based complementary logic on a Si substrate. *IEEE Electron Device Letters, 41*(6), 820–823. https://doi.org/10.1109/LED.2020.2987003.
90. Zheng, Z., et al. (2021). Monolithically integrated GaN ring oscillator based on high-performance complementary logic inverters. *IEEE Electron Device Letters, 42*(1), 26–29. https://doi.org/10.1109/LED.2020.3039264.
91. Amano, H., et al. (2018). The 2018 GaN power electronics roadmap. *Journal of Physics D: Applied Physics, 51*(16), 163001. https://doi.org/10.1088/1361-6463/aaaf9d.
92. Meneghesso, G., et al. (2006). Current collapse and high-electric-field reliability of un passi- vated GaN/AlGaN/GaN HEMTs. *IEEE Transactions on Electron Devices, 53*(12), 2932–2941. https://doi.org/10.1109/TED.2006.885681.
93. Stockman, A., et al. (2019). Threshold voltage instability mechanisms in p-GaN gate AlGaN/GaN HEMTs. In *2019 31st International Symposium on Power Semiconductor Devices and ICs (ISPSD)* (pp. 287–290). https://doi.org/10.1109/ISPSD.2019.8757667.
94. Hwang, I., et al. (2013). Impact of channel hot electrons on current collapse in AlGaN/GaN HEMTs. *IEEE Electron Device Letters, 34*(12), 1494–1496. https://doi.org/10.1109/LED.2013.2286173.
95. Li, X., et al. (2021). Investigating the current collapse mechanisms of p-GaN Gate HEMTs by different passivation dielectrics. *IEEE Transactions on Power Electronics, 36*(5), 4927–4930. https://doi.org/10.1109/TPEL.2020.3031680.
96. Tanaka, K., et al. (2015). Suppression of current collapse by hole injection from drain in a normally-off GaN-based hybrid-drain-embedded gate injection transistor. *Applied Physics Letters, 107*(16), 163502. https://doi.org/10.1063/1.4934184.
97. He, J., Tang, G., & Chen, K. J. (2018). VTH instability of *p*-GaN gate HEMTs under static and dynamic gate stress. *IEEE Electron Device Letters, 39*(10), 1576–1579. https://doi.org/10.1109/LED.2018.2867938.

98. Yang, S., et al. (2020). Identification of trap states in p-GaN layer of a p-GaN/AlGaN/GaN power HEMT structure by deep-level transient spectroscopy. *IEEE Electron Device Letters, 41*(5), 685–688. https://doi.org/10.1109/LED.2020.2980150.

99. Shi, Y., et al. (2019). Carrier transport mechanisms underlying the bidirectional V_{TH} shift in p-GaN gate HEMTs under forward gate stress. *IEEE Transactions on Electron Devices, 66*(2), 876–882. https://doi.org/10.1109/TED.2018.2883573.

100. Tang, X., et al. (2018). Mechanism of threshold voltage shift in *p*-GaN gate AlGaN/GaN transistors. *IEEE Electron Device Letters, 39*(8), 1145–1148. https://doi.org/10.1109/LED.2018. 2847669.

101. Sayadi, L., et al. (2018). Threshold voltage instability in p-GaNGate AlGaN/GaN HFETs. *IEEE Transactions on Electron Devices, 65*(6), 2454–2460. https://doi.org/10.1109/TED.2018. 2828702.

102. Sun, W., et al. (2019). Investigation of trap-induced threshold voltage instability in GaN-on-Si MISHEMTs. *IEEE Transactions on Electron Devices, 66*(2), 890–895. https://doi.org/10.1109/ TED.2018.2888840.

103. Joshi, R. P., et al. (2003). Analysis of dislocation scattering on electron mobility in GaN high electron mobility transistors. *Journal of Applied Physics, 93*(12), 10046–10052. https://doi.org/ 10.1063/1.1577406.

104. Prieur, E., et al. (1996). Threading dislocations in silicon layer produced by separation by implanted oxygen process. J*ournal of Applied Physics, 80*(4), 2113–2120. https://doi.org/10. 1063/1.363104.

105. Marino, F. A., et al. (2010). Effects of threading dislocations on AlGaN/GaN high-electron mobility transistors. *IEEE Transactions on Electron Devices, 57*(1), 353–360. https://doi.org/ 10.1109/TED.2009.2035024.

106. Hite, J. K., et al. (2014). Correlation of threading screw dislocation density to GaN 2-DEG mobility. *Electronics Letters, 50*(23), 1722–1724. https://doi.org/10.1049/el.2014.2401.

107. Duguay, S., et al. (2019). Evidence of Mg Segregation to Threading Dislocation in Normally-Off GaN-HEMT. *IEEE Transactions on Nanotechnology, 18*, 995–998. https://doi.org/10.1109/ TNANO.2019.2942400

108. Chen, J., et al. (2020). OFF-state drain-voltage-stress-induced VTH instability in Schottky-type p-GaN gate HEMTs. *IEEE Journal of Emerging and Selected Topics in Power Electronics,* 1–1. https://doi.org/10.1109/JESTPE.2020.3010408.

109. Zhou, G., et al. (2021). Determination of the gate breakdown mechanisms in p-GaN gate HEMTs by multiple-gate-sweep measurements. *IEEE Transactions on Electron Devices, 68*(4), 1518–1523. https://doi.org/10.1109/TED.2021.3057007.

110. Hu, C., & Chi, M.-H. (1982). Second breakdown of vertical power MOSFET's. *IEEE Transactions on Electron Devices, 29*(8), 1287–1293. https://doi.org/10.1109/T-ED.1982.20869.

111. Kuo, D. S., Hu, C., & Chi, M. H. (1983). dV/dt breakdown in power MOSFET's. *IEEE Electron Device Letters, 4*(1), 1–2. https://doi.org/10.1109/EDL.1983.25623.

112. Singh, P. (2004). Power MOSFET failure mechanisms. In *INTELEC 2004. 26th Annual International Telecommunications Energy Conference* (pp. 499–502). https://doi.org/10.1109/ INTLEC.2004.1401515.

113. Zhang, Z.-L., et al. (2017). Three-level gate drivers for eGaN HEMTs in resonant converters. *IEEE Transactions on Power Electronics, 32*(7), 5527–5538. https://doi.org/10.1109/TPEL. 2016.2606443.

114. Seidel, A., & Wicht, B. (2018). A fully integrated three-level 11.6nC gate driver supporting GaN gate injection transistors. In *2018 IEEE International Solid - State Circuits Conference - (ISSCC)* (pp. 384–386). https://doi.org/10.1109/ISSCC.2018.8310345.

115. Ke, X., et al. (2021). A 3-to-40-V automotive-use GaN driver with active bootstrap balancing and VSW dual-edge dead-time modulation techniques. *IEEE Journal of Solid-State Circuits, 56*(2), 521–530. https://doi.org/10.1109/JSSC.2020.3005794.

116. Huang, Q., et al. (2019). Predictive ZVS control with improved ZVS time margin and limited variable frequency range for a 99% efficient, 130- W/in3 MHz GaN totem-pole PFC rectifier. *IEEE Transactions on Power Electronics, 34*(7), 7079–7091. https://doi.org/10.1109/TPEL.2018.2877443.

117. Nitzsche, Maximilian, et al. (2020). Comprehensive Comparison of 99% Efficient Totem-Pole PFC with Fixed (PWM) or Variable (TCM) Switching Frequency. *PCIM Europe digital days 2020; International Exhibition and Conference for Power Electronics* (pp. 1–8). Renewable Energy and Energy Management: Intelligent Motion.

118. Chiu, H.-C., et al. (2015). Analysis of the back-gate effect in normally OFF p-GaN gate high-electron mobility transistor. *IEEE Transactions on Electron Devices, 62*(2), 507–511. https://doi.org/10.1109/TED.2014.2377747.

119. Li, X., et al. (2018). Suppression of the backgating effect of enhancement-mode p-GaN HEMTs on 200-mm GaN-on-SOI for monolithic integration. *IEEE Electron Device Letters, 39*(7), 999–1002. https://doi.org/10.1109/LED.2018.2833883.

120. Cree Wolfspeed. (Rev. 3, 07-2020). Silicon carbide power MOSFET C3M™MOSFET technology. In *C3M0060065J Datasheet*. https://www.wolfspeed.com/downloads/dl/file/id/1629/product/465/c3m0060065j.pdf.

121. Kaufmann, M., Seidel, A., & Wicht, B. (2020). Long, short, monolithic - the gate loop challenge for GaN drivers: Invited paper. In *2020 IEEE Custom Integrated Circuits Conference (CICC)*. Boston, MA (pp. 1–5).

122. Seidel, A., & Wicht, B. (2018). Integrated gate drivers based on high-voltage energy storing for GaN transistors. *IEEE Journal of Solid-State Circuits, 53*(12), 3446–3454. https://doi.org/10.1109/JSSC.2018.2866948.

123. Texas Instruments LMG3410 - Sample. (2017). url: https://www.systemplus.fr/wp-content/uploads/2017/07/SP17331_Texas_Instruments_LMG3410_600V_GaN_FET_Power_Stage_Sample_System_Plus_Consulting.pdf Retrieved from 21 June 2021.

124. Barbarini, E., & Radufe, N. (2017). Texas Instruments LMG3410 600V GaN FET Power Stage - Summary. https://www.systemplus.fr/wp-content/uploads/2017/07/SP17331_Texas_Instruments_LMG3410_600V_GaN_FET_Power_Stage_Flyer_System_Plus_Consulting-1.pdf. Retrieved from 21 June 2021.

125. STMicroelectronics. (2020). High power density 600V half bridge driver with two enhancement mode GaNHEMT. In *MASTERGAN1 Datasheet*. https://www.st.com/resource/en/datasheet/mastergan1.pdf.

126. Ayari, T., & El Gmili, Y. (2020). STMicroelectronics MASTERGAN1 half-bridge driver - sample. https://www.systemplus.fr/wp-content/uploads/2020/11/SP20580-STMicroelectronics-MASTERGAN1-Half-Bridge-Driver-Sample.pdf. Retrieved from 21 June 2021.

127. Navitas. (2020). GaNFast™power IC. In *NV6128 Datasheet*. https://www.navitassemi.com/download/nv6128-2/?wpdmdl=38977&ind=1611940435066.

128. Chen, H. -Y., et al. (2021). A fully integrated GaN-on-silicon gate driver and GaN switch with temperature-compensated fast turn-on technique for improving reliability. In: 2021 *IEEE International Solid- State Circuits Conference (ISSCC)* (pp. 460–462). https://doi.org/10.1109/ISSCC42613.2021.9365828.

129. Efficient Power Conversion. (Rev. 2.0) (2021). EPC2152 - 80 V, 15 A ePower™stage. In *Preliminary Datasheet*. https://epc-co.com/epc/Portals/0/epc/documents/datasheets/EPC2152_datasheet.pdf.

130. Navitas. (2021). Navitas produces world's first integrated half-bridge GaN power IC. https://www.navitassemi.com/navitas-producesworlds-first-integrated-half-bridge-gan-power-ic/. Retrieved from 12 April 2021.
131. Texas Instruments. (2016). UCC28880 700-V, 100-mA low quiescent current off-line converter. In *UCC28880 Datasheet*.
132. Power Integrations. (2016). Single-stage LED driver IC with combined PFC and constant current output for buck topology. In *LYT1402-1604 LYTSwitch-1 Family Datasheet*.
133. STMicroelectronics. (2014). Offline LED driver with primary-sensing and high power factor up to 15 W. In *HVLED815PF Datasheet*.
134. Wong, K., Chen, W., & Chen, K. J. (2010). Integrated voltage reference generator for GaN smart power chip technology. *IEEE Transactions on Electron Devices, 57*(4), 952–955. https://doi.org/10.1109/TED.2010.2041510.
135. Liao, C., et al. (2020). 3.8 A 23.6ppm/°C monolithically integrated GaN reference voltage design with temperature range from -50°C to 200°C and supply voltage range from 3.9 to 24V. In *2020 IEEE International Solid- State Circuits Conference - (ISSCC)* (pp. 72–74). https://doi.org/10.1109/ISSCC19947.2020.9062940.
136. Moench, S., et al. (2019). Integrated current sensing in GaN power ICs. In *2019 31st International Symposium on Power Semiconductor Devices and ICs (ISPSD)* (pp. 111–114). https://doi.org/10.1109/ISPSD.2019.8757678.
137. Basler, M., et al. (2020). A GaN-based current sense amplifier for GaN HEMTs with integrated current shunts. In *2020 32nd International Symposium on Power Semiconductor Devices and ICs (ISPSD)* (pp. 274–277). https://doi.org/10.1109/ISPSD46842.2020.9170047.
138. Kwan, A. M. H., et al. (2014). A highly linear integrated temperature sensor on a GaN smart power IC platform. *IEEE Transactions on Electron Devices, 61*(8), 2970–2976. https://doi.org/10.1109/TED.2014.2327386.
139. Li, X., Cui, M., & Liu, W. (2019). A full GaN-integrated sawtooth generator based on enhancement-mode AlGaN/GaN MIS-HEMT for GaN power converters. In *2019 International Conference on IC Design and Technology (ICICDT)* (pp. 1–3). https://doi.org/10.1109/ICICDT.2019.8790928.
140. Kwan, A. M. H., & Chen, K. J. (2013). A gate overdrive protection technique for improved reliability in AlGaN/GaN enhancement-mode HEMTs. *IEEE Electron Device Letters, 34*(1), 30–32. https://doi.org/10.1109/LED.2012.2224632.
141. Xu, H., et al. (2019). Integrated high-speed over-current protection circuit for GaN power transistors. In *2019 31st International Symposium on Power Semiconductor Devices and ICs (ISPSD)* (pp. 275–278). https://doi.org/10.1109/ISPSD.2019.8757685.
142. Tang, G., et al. (2017). Digital integrated circuits on an E-mode GaN power HEMT platform. *IEEE Electron Device Letters, 38*(9), 1282–1285. https://doi.org/10.1109/LED.2017.2725908.
143. Sun, R., et al. (2019). Development of GaN power IC platform and all GaN DC-DC buck converter IC. In *2019 31st International Symposium on Power Semiconductor Devices and ICs (ISPSD)* (pp. 271–274). https://doi.org/10.1109/ISPSD.2019.8757674.
144. Kaufmann, M., et al. (2020). 18.2 a monolithic E-mode GaN 15W 400V offline self-supplied hysteretic buck converter with 95.6% efficiency. In *2020 IEEE International Solid- State Circuits Conference - (ISSCC)*. San Francisco, CA (pp. 288–290).
145. Bush, S. (2020). ISSCC 2020: GaN power chip integrates control circuits. https://www.electronicsweekly.com/news/research-news/isscc-2020-gan-power-chip-integrates-control-circuits-2020-02/ Retrieved from 18 May 2021.

Circuit Integration in E-Mode GaN

3

In state-of-the-art e-mode GaN-on-Si process technologies, the HV power transistor is formed by a lateral structure. This allows for effortless integration of multiple components on one IC to extend the use of GaN from superior HV switching to additional functionality.

Figure 3.1 shows a cross section of the available devices in a typical e-mode GaN power process. The core component is the high-voltage power transistor (HV e-mode HEMT) depicted as first component on the left. On top of a silicon wafer used as a mechanical carrier, a GaN epitaxial layer is grown. To accommodate for the different crystal structures and thermal extension coefficients of silicon and GaN [1] a buffer layer is inserted as mechanical stress relief. On top of the GaN layer, AlGaN is deposited. Due to spontaneous piezoelectric polarization, electrons are generated at the interface surface and form a 2DEG as conductive channel of the transistor [2]. To control the conductivity of the channel, a metal gate is deposited above. A positive gate–source voltage V_{GS} accumulates more electrons in the channel, consequently reducing its resistance. In contrast, a negative V_{GS} depletes the channel and increases its resistance until the transistor is turned off and isolates. The electrons forming the transistor channel are generated by the crystal structure itself and, thereby, they are available for conduction even if no external gate–source voltage is applied. Thus, the native GaN transistor is a normally-on d-mode device. For safety requirements, normally-off e-mode transistors are preferred. This is achieved by adding p-doped GaN on top of the AlGaN. Thereby, the energy band structure of the AlGaN/GaN interface is changed and the 2DEG channel is depleted when the gate–source voltage V_{GS} is zero or even negative. A drain extension together with metal field plates is added to achieve a 650 V blocking capability for the transistor (see Chap. 2).

Using the layers and masks required for fabrication of this lateral HV e-mode HEMT, several other components can be formed as depicted in Fig. 3.1. By removing the drain extension and the field plates, a compact low-voltage transistor (LV e-mode HEMT) can be fabricated. It shows some characteristics which are typically similar to the HV HEMT, such as the threshold voltage V_{th}. Nevertheless, due to the shorter device length, it achieves

© The Author(s), under exclusive license to Springer Nature Switzerland AG 2022   53
M. P. Kaufmann and B. Wicht, *Monolithic Integration in E-Mode GaN Technology*,
Synthesis Lectures on Engineering, Science, and Technology,
https://doi.org/10.1007/978-3-031-15625-0_3

Fig. 3.1 Illustrated cross
section of components
available in a standard p-GaN
gate e-mode GaN process

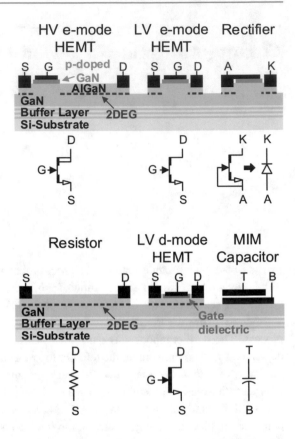

a higher transconductance g_m and smaller specific on-resistance R_{sp}. By shorting gate and source of the GaN HEMTs, a lateral field effect rectifier (LFER) can be implemented [3] accordingly to a NMOS diode in standard silicon processes. To implement this rectifier, both HV and LV HEMT can be used. This leads to similar voltage ratings for the rectifiers as for the transistors.

If the transistor gate including the p-GaN is removed, the resulting structure implements a resistor. It is formed by the normally-on 2DEG channel with a specific sheet resistance of around 500 Ω/□ [4]. With an isolated gate electrode above the 2DEG channel, the conductivity of the device can be modulated. With a positive gate–source voltage, more electrons are accumulated to increase the channel conductivity. A negative V_{GS} reduces the electrons in the channel until it is fully depleted. This native GaN device is a LV d-mode HEMT. It can be implemented using only the available layers. However, designated masks and process steps are required in order to achieve a well-controlled threshold voltage in the range of -2 V and low leakage currents for the depletion mode device. Therefore, it is not offered by default in every e-mode GaN process. The available component set is concluded by MIM capacitors. Depending on the utilized metal layers and spacing between the conductive metal, capacitors with different breakdown voltages can be fabricated. Typically, capacitance densities in the range of 1 fF/μm^2 can be achieved for 20 V rated capacitors (estimated from [5]).

The lack of suitable p-type devices (see Sect. 2.2) poses some challenges for circuit integration in GaN. On the one hand, this prevents the use of established CMOS techniques for logic gates and gate drivers. On the other hand, circuits like comparators and operational amplifiers with low input common mode levels can not be implemented by simply using a p-type input stage. However, the available components can be combined to integrate various circuits. The following sections explore the characteristics and properties of the GaN process for analog, digital, mixed-signal, and high-voltage integration. Challenges posed by the immature process technology as well as the lack of p-type devices are discussed and different circuit techniques to address these challenges are proposed and investigated. This forms the base for higher levels of integration as demonstrated with an integrated a power converter IC in Chap. 4.

3.1 Analog Integration

With the available components in the GaN process (Fig. 3.1), different circuits and functions can be designed and integrated in one IC. The most important components for analog integration are the LV HEMT and the 2DEG resistor to form amplifier stages. They are supported with MIM capacitors and rectifiers which enable bootstrapping techniques.

The intrinsic device physics are different for transistors in GaN and silicon CMOS technology (see Chap. 2). Despite that, they can be handled similarly from a design perspective. GaN transistors have three main terminals, gate, source, and drain. Similar to silicon transistors, they can be operated in different operation areas (linear region, saturation region, subthreshold, moderate inversion, and strong inversion). The behavior of the transistors can be linearized around an operating point with the established small-signal parameters, e.g. transconductance g_m, output impedance r_{ds}, capacitances c_{gs}, and others.

A fundamental small-signal model of a GaN HEMT is depicted in Fig. 3.2a. It is based on the established small-signal model for a silicon NMOS transistor, which is also applicable for GaN HEMTs [6]. In contrast to the silicon NMOS transistor, the small-signal gate resistances r_{gs} and r_{gd} are not related to oxide leakage currents but to the back-to-back diode configuration of the p-GaN gate structure, Fig. 3.2b [7]. Thus, r_{gs} and r_{gd} show a strong dependence on the DC-bias voltage as well as the temperature. In the following, important analog properties such as intrinsic gain, achievable single-stage gain, matching properties, and noise of transistors are characterized and compared to silicon technology. The implications of the analog properties of GaN devices on circuit design are discussed and n-type only design techniques are presented.

Intrinsic Gain and Single-Stage Gain

The achievable gain in a single amplifier stage is a key parameter of a semiconductor technology and can therefore be used as one parameter to compare analog properties and per-

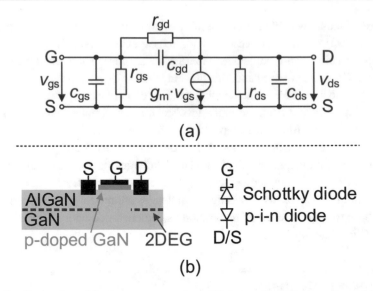

Fig. 3.2 **a** Small-signal model of a n-type transistor [6]; **b** transistor cross section with equivalent gate model specific for p-GaN gate GaN transistors [7]

formance of different technologies. For many functions, especially analog control loops, amplifiers are used in a negative feedback configuration where the amplifier gain defines the permanent control deviation. If high accuracy is required, amplifiers with a voltage gain of 100 dB and more are desired. This can be achieved using any technology by connecting multiple amplifier stages in series. However, each stage adds a pole to the transfer function of the full amplifier. This leads to increased difficulties for ensuring stability of the complete amplifier. Therefore, a high single-stage gain is preferable in order to achieve a defined amplification with a small amount of stages. The following part first characterized the intrinsic gain of GaN transistors and compares it with silicon technologies. After that, the implications of the absence of suitable complementary devices in GaN (see Sect. 2.2) on the achievable single-stage gain are investigated. A circuit to improve the gain of GaN amplifiers without utilizing p-type or d-mode transistors is presented and characterized before the effects of a d-mode device on the single-stage gain are discussed.

Intrinsic Gain of GaN Transistors

The theoretical maximum voltage gain of a semiconductor technology is the intrinsic voltage gain A_{int} of a transistor. It is defined as the product of the small-signal transconductance g_m and the output resistance r_{ds}. In a first-order approximation, g_m is proportional to the transistor width W and r_{ds} is proportional to $1/W$. Therefore, the normalized values g_m/W and $r_{ds} \cdot W$ for a defined transistor length L are general technology parameters, which can be used to obtain the general expression for the intrinsic transistor gain A_{int} given in Eq. 3.1.

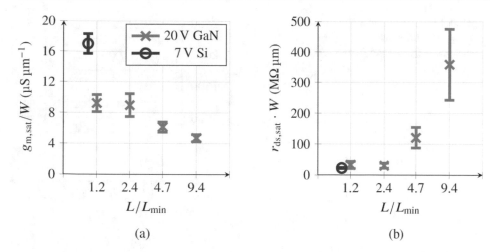

Fig. 3.3 **a** Measured $g_{m,sat}$ and **b** measured $r_{ds,sat}$ in saturation region for differently sized transistors at $I_D = 1\,\mu A/\mu m$

$$A_{int} = \frac{g_m}{W} \cdot r_{ds} \cdot W = g_m \cdot r_{ds} \tag{3.1}$$

Figure 3.3a shows the characterized values for $g_{m,sat}$ of several, differently sized LV GaN transistors. $g_{m,sat}$ is the transconductance g_m of a transistor operated in saturation region with $V_{DS} > V_{GS} - V_{th}$. The corresponding $r_{ds,sat}$ (small-signal drain–source resistance r_{ds} in saturation region) for the same transistors is depicted in Fig. 3.3b. For all values in Fig. 3.3, the measurements are conducted at a drain current density of $1\,\mu A/\mu m$ transistor width. Twenty-seven samples across one wafer are characterized for each transistor size. The marker x indicates the average value and the error bar shows the standard deviation of the 27 measured values. As a reference, the values for a LV silicon transistor in a HV silicon technology are added with the red o to the plots. While GaN transistors show a slightly smaller g_m than silicon devices, the r_{ds} is marginally higher.

Transistors with a constant width show a larger channel resistance for an increasing device length L. This leads to a reduced drain current at the same conditions for V_{GS} and V_{DS}. Therefore, g_m decreases for increasing L and at the same time r_{ds} increases with larger L. This expected behavior is verified by the measured results shown in Fig. 3.3.

Based on Eq. 3.1, the intrinsic gain can be calculated for each characterized transistor by multiplying the value of g_m from Fig. 3.3a with the respective r_{ds} of Fig. 3.3b. The result is depicted as A_{int} in Fig. 3.4. The intrinsic voltage gain of the examined short channel transistors is similar for both, silicon and GaN technology at around 300 (50 dB). It increases strongly with the channel length of the devices to values above 1000 (60 dB) for $L = 9.4 \cdot L_{min}$, as measured for differently sized GaN transistors. This is expected from the behavior of g_m and r_{ds} over channel length in Fig. 3.3 where r_{ds} increases much more than g_m decreases. However, all values of g_m, r_{ds}, and A_{int} for the GaN transistors show a large

Fig. 3.4 Measured intrinsic
gain A_{int} for different sized
transistors at $I_D = 1\ \mu\text{A}/\mu\text{m}$

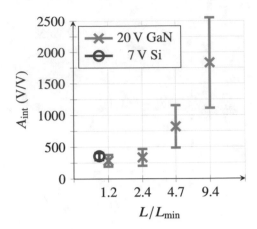

spread of around $\pm 50\%$, especially for long devices which is likely caused by imperfections of the GaN epitaxial layer (see Sect. 2.2). The implications of the large spread on circuit design will be discussed later in this section.

The values for A_{int} displayed in Fig. 3.4 are for a drain current density of $1\ \mu\text{A}/\mu\text{m}$ transistor width. At higher current, g_m of the transistors increases, while r_{ds} reduces. Thus, the values of A_{int} show only a minor variation with the drain current which is much smaller than the transistor-to-transistor variation on the examined wafer.

Single-Stage Gain with Resistor Load

The transistor operated in saturation region acts as transconductance amplifier, where a voltage signal V_{GS} at the gate is translated into a current signal I_D. A load has to be added to the drain current path in order to translate the current signal back to an amplified voltage signal and thereby forming a basic voltage amplifier stage. This load also provides the bias current for the amplifier. In silicon CMOS, this is typically done utilizing a PMOS transistor in current source configuration. A defined bias voltage V_{bias} is applied to the gate of the PMOS device and it is operated in saturation region. Thereby, a bias current is provided with the high output impedance $r_{\text{ds,sat}}$ of the transistor and shows therefore a current source characteristic. However, in GaN technology, a p-type device is typically not available (see Sect. 2.2). Hence, a resistor is used as load to provide the bias current and perform the current-to-voltage conversion.

Figure 3.5a shows the schematic of a common-source amplifier in GaN technology using a resistor R_{bias} to provide a DC-bias current for the transistor. A typical silicon implementation of a common-source amplifier with a PMOS current source as load is depicted in Fig. 3.5b. For the small-signal model in Fig. 3.5c, both the supply voltage rail as well as the reference potential are shorted to AC-ground. The impedance r_{load} represents the small-signal impedance of either the bias resistor in GaN or the PMOS current source in silicon CMOS technology. Based on this model and Eq. 3.1, the voltage gain A_V of the common-source amplifiers can be calculated as defined in Eq. 3.2:

Fig. 3.5 Schematics of common-source amplifiers with **a** resistor load in a GaN process, **b** PMOS current source load in a silicon process and **c** generic small-signal model of a common-source amplifier

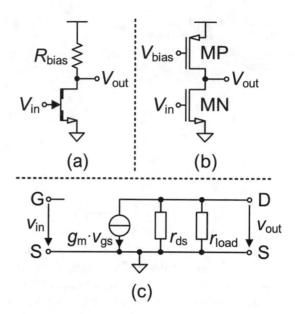

$$A_V = \frac{v_{out}}{v_{in}} = g_m \cdot \frac{r_{ds} \cdot r_{load}}{r_{ds} + r_{load}} \tag{3.2}$$

Assuming similar r_{ds} for both the PMOS and the NMOS transistors in a silicon process, the typical voltage gain A_V of a common-source amplifier is half of the intrinsic gain A_{int}. If, instead, a bias resistor is used, this changes quite a lot. Assuming a typical supply voltage $V_{DD} = 6$ V for GaN transistors and a DC output voltage of the amplifier $V_{out} = V_{DD}/2$, the resistor has to provide the required bias current with a voltage drop of 3 V. To generate a bias current $I_D = 1$ μA, the resistance can be calculated as given in Eq. 3.3.

$$R_{bias} = \frac{V_{DD} - V_{out}}{I_D} = \frac{3 \text{ V}}{1 \text{ μA}} = 3 \text{ MΩ} \tag{3.3}$$

The average r_{ds} of a short-channel GaN transistor with $L = 1.2 \cdot L_{min}$ is characterized to be approximately 30 MΩ, see Fig. 3.3b. This is ten times larger than the value for $R_{bias} = 3$ MΩ calculated in Eq. 3.3. For a linear resistor both the small-signal and the large-signal resistance are the same $r_{load} = R_{bias} \sim r_{ds}/10$ in Eq. 3.2. Thereby, the voltage gain A_V reduces to less than one tenth of the intrinsic gain A_{int} for a common-source amplifier formed by a GaN transistor with a resistor load. Thus, the achievable single-stage gain in the GaN process is around 28 linearly (equals 29 dB), while, in a silicon CMOS technology, a linear voltage gain of 130 (equals 42 dB) can be achieved. This is estimated as half of the intrinsic gain for a silicon NMOS transistor (see Fig. 3.4).

Various, differently sized common-source amplifiers are characterized across one GaN wafer in order to verify the assumptions above. The measured voltage gain A_V of the

Fig. 3.6 Measured voltage
gain A_V of common-source
amplifiers for different bias
currents

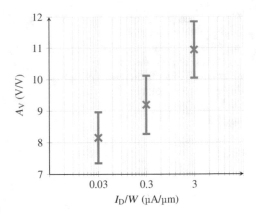

common-source amplifiers is shown in Fig. 3.6. Over a bias current range of 0.03 to 3
$\mu A/\mu m$ transistor width the linear voltage gain is only between 8 and 11. While this is a bit
lower than estimated, it is within the same order of magnitude. Furthermore, it confirms that
the achievable single-stage gain of a basic common-source amplifier in GaN technology
is much lower than in CMOS technology. This is mainly due to the resistor biasing the
amplifier instead of a PMOS current source as it is common in silicon technologies.

Gain Boosting without Complementary Transistors

In order to enhance the voltage gain of a GaN common-source amplifier without employing
a p-type device, an n-type transistor can be used as current source instead. It can provide the
required bias current with a higher small-signal output resistance r_{load} than a simple resistor.
A constant V_{GS} has to be applied to the n-type transistor in order to achieve a current source
characteristic. Since the source terminal of an n-type load transistor is connected to the
amplifier output V_{out}, the gate voltage has to be provided with respect to V_{out}. Thus, a
bootstrapping technique is implemented to apply a constant V_{GS} independent of the source
potential.

One possible implementation of such a common-source amplifier with a bootstrapped
n-type transistor as load is illustrated in Fig. 3.7a. Transistor Q_A is utilized as amplifier
while Q_{CS} provides the bias current and acts as load. Figure 3.7b shows the biasing phase
when the control signal $\phi = \text{'1'}$. During this phase, the output V_{out} is at ground potential,
while the gate and the drain of Q_{CS} are connected together to form a NMOS-diode. A
source degeneration resistor R_S is added to Q_{CS} as a negative feedback in order to reduce
the effective transconductance and increase the output resistance. This stabilizes I_{bias} by
compensating variations of V_{bias}, which would otherwise directly affect I_{bias} through the g_m
of the transistor Q_{CS}. A more detailed discussion of such a resistive source degeneration is
provided further below.

The resistor R_{bias} is required to set V_{bias} to a value smaller than V_{DD}. This provides
enough headroom for the bias current source circuit to ensure operation in saturation region

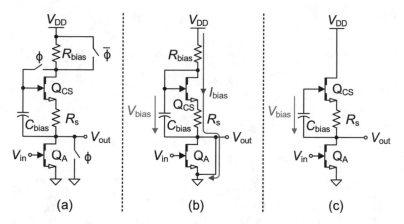

Fig. 3.7 **a** Schematic of a common-source amplifier with bootstrapped n-type current source load, **b** resulting schematic for $\phi = $ '1' and $\bar{\phi} = $ '0', **c** resulting schematic for $\phi = $ '0' and $\bar{\phi} = $ '1'

for Q_{CS}. With this circuit, the resulting bias current can be calculated as given in the relation in Eq. 3.4:

$$I_{bias} = \frac{V_{DD}}{R_{bias} + R_S + R_{ds,Qcs}} \tag{3.4}$$

During this phase, $V_{bias} \sim (V_{th} + I_{bias} \cdot R_S)$ is stored on the capacitor C_{bias}. During the other phase when $\bar{\phi} = $ '1', the circuit acts as amplifier. This is illustrated in Fig. 3.7c. C_{bias} keeps the gate voltage of Q_{CS} constant. Hence, it acts as a current source to provide the bias current for the amplifier with a high output resistance. Without R_S, any change of V_{bias} would directly affect I_{bias} through the transconductance g_m of the transistor. A small drop of V_{bias} due to leakage of the integrated capacitor could significantly reduce I_{bias} and, thus, lead to a complete de-biasing of the amplifier. As one countermeasure, the length of Q_{CS} is chosen to $7 \cdot L_{min}$ to reduce its g_m (see also Fig. 3.3). Additionally, the source degeneration implemented by R_S further reduces the effective transconductance. Therefore, R_S is sized to cause a voltage drop $R_S \cdot I_{bias} \sim 0.5 \cdot V_{th,Qcs}$.

In order to prove the functionality and to validate the gain-boosting performance of the proposed circuit, transient and DC measurements are conducted. Therefore, two different integrated common-source amplifiers are characterized. Both of them are designed using same-sized amplifier transistors (Q_A in Fig. 3.7) biased with the same DC drain current I_{bias}. However, one is designed with the proposed bootstrapped n-type transistor as current source according to Fig. 3.7, while the other is implemented with a simple resistor load (Fig. 3.5a).

Figure 3.8a shows the measured transient waveforms of V_{in} and V_{out} for the common-source amplifier with bootstrapped current source load. The input signal V_{in} is a sine wave with an amplitude of 50 mV and a frequency of 10 kHz at a DC voltage of 1.8 V. Since the common-source amplifier is an inverting amplifier, the sine wave of V_{out} is in opposite polar-

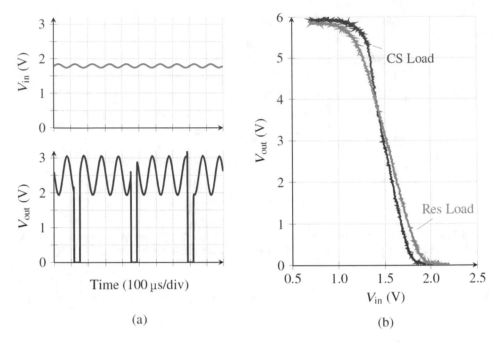

Fig. 3.8 Measured results of a common-source amplifier with bootstrapped current source load: **a** transient waveforms, **b** DC characteristics compared to an amplifier with resistor load

ity with respect to V_{in} and shows an amplified amplitude of 550 mV. In periodic intervals, V_{out} is pulled to ground when the bootstrapping clock $\phi = $ '1'. During this time, the voltage V_{bias} on capacitor C_{bias} is recharged to compensate for discharge losses mainly caused by gate leakage. The circuit is characterized and operates properly for a wide frequency range of the bootstrapping clock between 500 Hz and 200 kHz. When the amplifier is implemented in a clocked system such as a switched-mode power converter, the system frequency can also be utilized as bootstrapping clock to recharge C_{bias} when the amplifier is not needed. In continuous-time systems, the frequency of the bootstrapping clock may be chosen to a value far greater than the signal bandwidth to obtain a quasi-time-continuous amplifier with low distortion.

A comparison of the DC characteristics V_{out} over V_{in} between a common-source amplifier with resistor load and the presented bootstrapped current source load is depicted in Fig. 3.8b. Both amplifiers are biased with the same current $I_{bias} = 0.25\ \mu A/\mu m$ at $V_{out} = 3$ V. The steeper slope of the amplifier with bootstrapped current source load ("CS Load" in Fig. 3.8b) points to a higher voltage gain. This indicates that I_{bias} is, as desired, provided with a higher small-signal load resistance r_{load}.

In the next step, the frequency behavior of both amplifiers is characterized. The voltage gain over frequency for both amplifiers is shown in the bode plot in Fig. 3.9. The amplifier with bootstrapped current source load shows a 5 dB (1.8x) higher DC gain and a higher

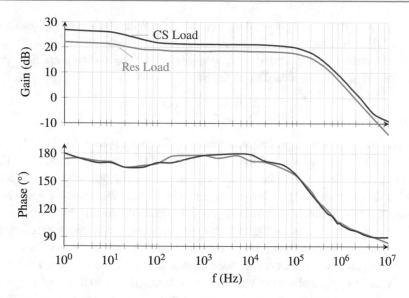

Fig. 3.9 Bode plot of common-source amplifiers with resistor load and bootstrapped current source load

voltage gain over the full frequency range. Additionally, the unity gain frequency as well as the the 3 dB cutoff frequency is higher, while the phase characteristic is similar for both amplifiers.

For both amplifiers, the channel resistance of the transistors and also the bias resistor show a strong temperature coefficient of around 12,000 ppm/K typical for the 2DEG in GaN (see Sect. 2.1). Towards higher temperature, all resistances increase, while the capacitances stay approximately constant. This leads to a bandwidth degradation of amplifier stages at higher temperatures. This is also confirmed by the investigations presented in [8], where the bandwidth reduces by more than 50% when the temperature changes from room temperature of 27 °C to 200 °C. In the same temperature range, the DC gain reduces also by ~50% mainly caused by lower bias currents at higher temperatures (see also Fig. 3.6).

While the amplifier with bootstrapped current source shows a higher voltage gain than the amplifier with a resistor pull-up, its gain is still significantly lower than the voltage gain of a CMOS amplifier utilizing a PMOS current source. The main reasons for that are the parasitic bottom- and top-plate capacitances of the integrated C_{bias}. During charging phase, while $\phi = $ '1', the bottom plate of C_{bias} is at ground and the top plate is charged to $V_{bias} = V_{th} + I_{bias} \cdot R_S$. When ϕ changes to '0' and $\bar{\phi}$ to '1', the bottom plate of C_{bias} is pushed to V_{out} and the top plate is floating together with the gate of Q_{CS}. Due to the parasitic capacitance between the top plate of C_{bias} and the grounded substrate, the voltage V_{bias} reduces by a fraction of V_{out}, which is defined by the capacitance ratio between C_{bias} and its parasitic top plate capacitance. For operation as class A amplifier, the DC operating point is at $V_{out} = V_{DD}/2$. When the circuit changes from C_{bias} recharging phase to amplification mode, V_{out} rises from ground potential to $V_{\text{DD}}/2 = 3$ V. With a typical capacitance ratio for integrated MIM capacitors $C_{\text{top}}/C_{\text{bias}} = 10\%$, this leads to a reduction of V_{bias} by 300 mV, consequently reducing I_{bias}. Furthermore, any change of V_{out} during amplifier operation also changes V_{bias} and I_{bias} according to Eq. 3.5.

$$\Delta I_{\text{bias}} \sim \frac{\Delta V_{\text{bias}}}{R_{\text{S}}} = -\frac{0.1 \cdot \Delta V_{\text{out}}}{R_{\text{S}}} \tag{3.5}$$

The parasitic top plate capacitance therefore leads to a negative feedback loop in the amplifier. A rise of V_{in} reduces the $R_{\text{DS,on}}$ of transistor Q_{CS} and, therefore, pulls V_{out} to a lower value. Due to the bottom-plate effect V_{bias} rises, which in turn causes I_{bias} to increase. The higher I_{bias} consequently leads to a larger voltage drop at the channel resistance of the transistor used as amplifier. This causes V_{out} to rise, which counteracts the original amplifier operation mechanism. Therefore, the amplifier with bootstrapped current source load is only able to work properly, when the negative feedback caused by C_{top} is reduced. For this reason, a large enough R_{S} is utilized as source degeneration and the current source transistor Q_{CS} is designed as long channel device with low g_{m} reducing the dependence of I_{bias} on V_{bias}. The achievable voltage gain with such a circuit could also be improved by a smaller ratio $C_{\text{top}}/C_{\text{bias}}$. Simulations suggest that for a bottom plate capacitance of 5% instead of 10%, a single-stage gain above 40 dB (100 linear) can be achieved. To summarize, the amplifier with bootstrapped current source load achieves higher voltage gain than the one with resistor load, but it comes with some limitations:

- The presented common-source amplifier is a clocked system which supports time-continuous operation only for low signal frequencies below 100 kHz where sufficient oversampling can be achieved.
- Due to the parasitic top plate capacitance of the integrated capacitor C_{bias}, V_{bias} is not constant but depends on V_{out}. This leads to a negative feedback loop. Hence, the achievable gain of the proposed circuit is lower than for a silicon amplifier with a p-type current source load.
- Some of the switches required for the bootstrapped current source circuit have to be turned on while guiding voltages as high as the general supply voltage V_{DD}. To enable this without p-type devices, a bootstrapped switch has to be implemented. An example for such a bootstrapped switch is presented in Sect. 3.3.

Another way to provide the bias current for a common-source amplifier with a high small-signal impedance is to employ a d-mode transistor as current source. Figure 3.10 depicts a schematic of such a common-source amplifier with d-mode current source load. The d-mode current source provides a current I_{bias} based on the threshold voltage of the transistor and the resistance value as given by Eq. 3.6.

$$I_{\text{bias}} = \frac{V_{\text{GS},Q_{\text{d}}}}{R_{\text{bias}}} \sim \frac{V_{\text{th},Q_{\text{d}}}}{R_{\text{bias}}} \qquad \text{for} \quad \frac{I_{\text{bias}}}{g_{\text{m}}} \ll V_{\text{th},Q_{\text{d}}} \tag{3.6}$$

Fig. 3.10 Schematic of a commons source amplifier with d-mode current source load

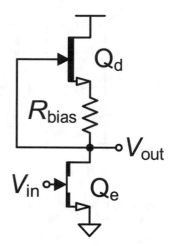

When the d-mode transistor is operated at low current density, the resulting I_{bias} depends only on R_{bias} and V_{th}. Since there is no influence of g_m, also variations of g_m are compensated. This will be discussed in detail for a source-degenerated current mirror (see Fig. 3.16).

While the d-mode current source provides the bias current with a high output impedance and does not suffer from parasitic bottom plate effects of integrated capacitors, it fundamentally limits the output voltage range. For proper current source operation with a high output impedance, the d-mode transistor has to be operated in saturation region. Therefore, the d-mode current source requires a voltage drop of at least V_{th,Q_d}, leading to a limit for the output voltage $V_{out} < V_{DD} - V_{th,Q_d}$.

Since the fabrication process steps of d-mode and e-mode GaN transistors are very similar, some technologies offer both devices in one process flow [3]. However, additional masks are required to fabricate proper d-mode devices with well-controlled parameters. Therefore, they are not available in all GaN processes and this implementation of a common-source amplifier is limited to special GaN technologies.

GaN transistors achieve an intrinsic gain similar to silicon transistors. However, the lack of suitable p-type devices fundamentally limits the achievable single-stage gain of amplifier circuits. A circuit to improve the gain of GaN amplifiers by employing a bootstrapped n-type transistor as current source load is proposed. However, due to the parasitic top-plate capacitance of integrated capacitors, the benefit of the circuit is limited.

Initial characterization results for g_m and r_{ds} of integrated GaN transistors show large parameter variations. The implications of these variations on the matching of devices in that technology are investigated in the following part.

Matching of Resistors and Transistors

In state-of-the-art semiconductor technologies, absolute parameters of components generally show large variations. The parameters vary considerably on one wafer and even more when the scope, in which the parameter is tracked, increases to wafer lots or even to different

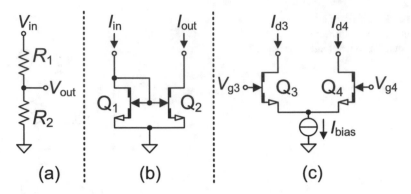

Fig. 3.11 Fundamental analog circuits relying on device matching: **a** resistor voltage divider, **b** current mirror, and **c** differential pair

fabrication runs. Typical parameters of semiconductor technologies such as the sheet resistance (R_{sp}), the threshold voltage (V_{th}) and the transconductance (g_m) may show a variation of $\pm 30\%$ and higher [9] when single devices are compared over different fabrication lots. However, the closer the two devices are located, the more similar their parameters become. If the devices are fabricated in the same wafer lot, the parameter variation is already lower. If the devices are located on the same wafer or even close to each other in direct proximity, the difference of parameters between these devices can be reduced to less than $\pm 5\%$. Therefore, most integrated circuits do not rely on absolute values but rather on relative relations. Additionally, several layout techniques such as utilizing unit devices, adding dummy structures for same neighborhood, and arranging devices in a common centroid style with same current direction are employed to further improve the matching between adjacent devices [10]. These techniques are also employed for the GaN layouts of this work used to investigate the matching of resistors and transistors in this technology.

To obtain knowledge about matching properties of the utilized GaN process, basic analog test structures (see Fig. 3.11) are characterized. The results are used to extract device properties and matching characteristics of various device parameters such as resistance, transconductance g_m, and threshold voltage V_{th}.

Matching of Resistors

Fig. 3.11a shows a resistor divider as one example circuit for matching of resistances. It generates a voltage V_{out} based on an input voltage V_{in} according to Eq. 3.7. V_{out} is independent of the absolute values for R_1 and R_2 but depends only on the ratio between the resistance values.

$$V_{out} = V_{in} \cdot \frac{R_2}{R_1 + R_2} \tag{3.7}$$

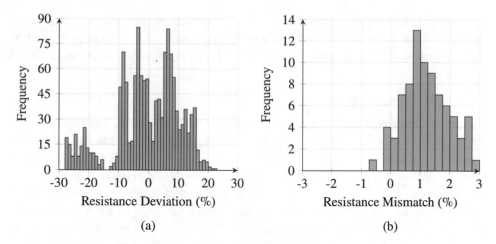

Fig. 3.12 **a** Resistance deviation of 1364 similar resistors spread over one wafer showing a standard deviation $\sigma = 10.9\%$, **b** mismatch of 82 resistor pairs with interdigitated layout showing a systematic mismatch $\mu = 1.2\%$ and a standard deviation $\sigma = 0.76\%$

In order to characterize the resistance variation over one wafer as well as to extract the matching properties of adjacent resistors, various I–V curves are measured for ~ 1400 resistors distributed across one wafer. All resistors are composed of the same number of squares and, hence, should show the same resistance value.

In the first step, the absolute resistance variations are investigated for the utilized GaN process. Thus, the resistance value of each resistor is extracted from the measured I–V characteristics. Additionally, the deviation of each resistance value from the average value of all characterized resistors is calculated. Figure 3.12a shows a histogram of these relative resistance deviations $(R/R_{mean} - 1) \cdot 100\%$ for 1364 characterized resistors. The absolute values of the resistors distributed all across the wafer show a standard deviation $\sigma = 10.9\%$.

In the second step, the matching of adjacent resistors is examined. Figure 3.12b shows the variation of 82 resistor pairs distributed across the same wafer. These resistor dividers are designed for good matching with interdigitated layout style based on unit cells. Both resistors of each matched pair are designed for the same resistance value $R_1 = R_2$. The mismatch between the two resistors is calculated as $(R_1/R_2 - 1) \cdot 100\%$. The mismatch shows a standard deviation of $\sigma = 0.76\%$. As expected, this value is more than ten times lower than the standard variation of the absolute resistance values previously characterized. However, the mismatch of R_1 and R_2 shows a systematic offset of $\mu = 1.2\%$, meaning that R_1 is typically larger than R_2. This is caused by the backgating effect in the GaN process. The silicon carrier wafer used as substrate is connected to ground, acting as backgate for the resistor (see Fig. 3.1). Since V_{in} and V_{out} are positive voltages, this leads to a negative backgate voltage V_{BS} for the resistors, reducing the 2DEG concentration in the resistor channel. R_1 is connected between V_{in} and V_{out} and thereby experiences a larger backgate voltage V_{BS} compared to R_2. This leads to a systematically larger resistance value for R_1 despite the same geometries for both resistors. Nevertheless, such a systematic effect can be compensated by adapting the resistance values of R_1 and R_2 according to the voltage levels of V_{in} and V_{out}.

Current Matching of Transistors

The initial characterization results for g_m and r_{ds} of GaN transistors (see Fig. 3.3) show large variations of these absolute parameters for samples distributed across one wafer. For resistors it has been demonstrated, that a large variation of the absolute sheet resistance in the order of $\sigma = 10\%$ can be reduced to values below $\sigma = 1\%$ for a matched pair of adjacent devices. Hence, the parameter variation of adjacent matched transistors is investigated. A current mirror (see Fig. 3.11b) is utilized as investigation vehicle for a basic analog circuit relying on the matching of relative parameters. It generates a current I_{out} based on an input current I_{in} according to Eq. 3.8. The current matching between I_{in} and I_{out} relies on the similarity of two transistors in order to provide the same drain current I_D at the same gate-source voltage V_{GS}. Therefore, the current mirror requires good matching for both, V_{th} and g_m of the two transistors. After a theoretical analysis of the influence of g_m- and V_{th}-mismatch on the current transfer function of a current mirror, the mismatch of V_{th} and g_m between matched transistors is characterized separately for various current mirror samples. The characterized mismatch of GaN transistors is compared to published data of silicon technologies. Additionally, source degeneration as a method for compensating g_m mismatch on circuit level is discussed.

According to Eq. 3.8, mismatch of g_m or V_{th} have a different influence on the current transfer function of the basic current mirror. Mismatch of g_m of the two transistors leads to a gain error. This is illustrated in Fig. 3.13a where, depending on the g_m ratio, the curves I_{out} over I_{in} show different slopes. The influences of V_{th} mismatch of the transistors are more complicated, since the common gate-source voltage V_{GS} increases with higher I_{in}.

$$I_{out}\,(I_{in}) = \frac{g_{m2}}{g_{m1}} \cdot \frac{V_{GS}(I_{in}) - V_{th2}}{V_{GS}(I_{in}) - V_{th1}} \cdot I_{in} \tag{3.8}$$

At higher current levels when V_{GS} is considerably larger than V_{th}, mismatch of V_{th} mostly leads to an offset error for the output current. This can be seen in the parallel curves in Fig. 3.13b for input currents above $\sim 30\,\mu A$ when V_{GS} is significantly larger than V_{th}. For small current levels below $10\,\mu A$, a V_{th} mismatch also translates into a gain error for the current mirror. This leads to different slopes for I_{out} over I_{in} at low current values (Fig. 3.13b).

As derived above, the standard deviation of the drain currents for two adjacent transistors shows dependencies on both, V_{th} and g_m matching [11]. In the first step, the variations and matching of V_{th} are investigated for differently sized transistors. Figure 3.14 shows the V_{th} matching plot as proposed by Pelgrom as standard deviation σ of measured threshold voltage differences ΔV_{th} of matched transistor pairs. The values for silicon are extracted

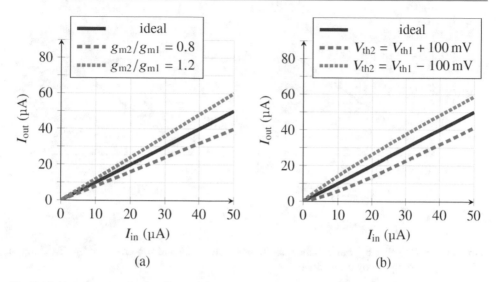

Fig. 3.13 Calculated current mirror transfer function for **a** g_{m} mismatch and **b** V_{th} mismatch

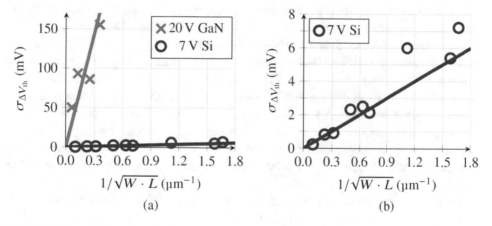

Fig. 3.14 V_{th} matching plot after [12] **a** GaN and Silicon comparison **b** zoom-in on Silicon (drawn after [12])

from [12]. As generally known for silicon technologies, the matching improves with larger device area following the square root law $\sigma_{\Delta V_{\mathrm{th}}} \propto 1/\sqrt{W \cdot L}$.

In order to confirm if this characteristic is also true for GaN transistors, the ΔV_{th} is characterized for adjacent GaN transistors with differently sized gates. To calculate the standard deviation, 27 pairs of matched transistors are characterized individually for each gate area size. For best matching, all transistor pairs are laid out in a common centroid style and show equal current direction. Additional dummies are added to achieve the same neighborhood for all transistors. The results of the V_{th} mismatch characterization are plotted

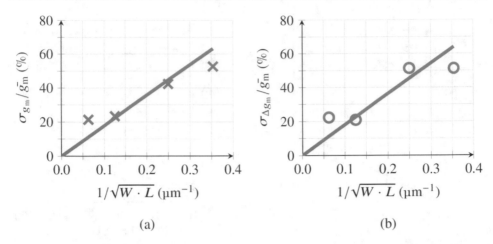

Fig. 3.15 **a** Normalized g_m distribution across one wafer GaN and **b** normalized g_m matching of adjacent transistor pairs over gate area

with x markers in Fig. 3.14a. Furthermore, a linear regression curve is added similar to the method proposed for silicon in [12]. Both the silicon and the GaN technology show a similar dependence of the mismatch related to the device size $\sigma_{\Delta V_{th}} \propto 1/\sqrt{W \cdot L}$. However, the V_{th} mismatch of GaN transistors is nearly two decades higher than for similarly sized silicon transistors. This fits to published V_{th} uniformity data: [13] shows standard deviations $\sigma_{V_{th}} = 229$ mV for large, discrete GaN transistors which benefit from averaging effects due to much larger device areas. A wafer shot map published in [14] shows $\sigma_{V_{th}} = 49$ mV.

In order to achieve integrated circuits in GaN technology with good accuracy, sensitive transistors need to be designed with a large gate area. Additionally, dedicated circuitry should be implemented to compensate for device variations as discussed in the following and in Sect. 4.3.

After the characterization of V_{th} matching, the variations of the transconduction g_m is examined. Figure 3.15a shows the standard deviation of g_m normalized to the average of g_m as function of the transistor gate area of the same transistors characterized for V_{th} mismatch in Fig. 3.14. Figure 3.15b shows the g_m matching of neighboring transistors normalized to the average of g_m.

Similar to the mismatch of V_{th}, a linear relation $\sigma_{\Delta g_m}/\bar{g_m} \propto 1/\sqrt{W \cdot L}$ can be observed. However, even for the transistor pairs with the largest gate area characterized in this measurement series, the g_m mismatch is above 20%. Thus, even larger transistors have to be used in order to design precise circuits. As an alternative option, additional circuitry may be implemented to compensate for the mismatch of g_m.

In order to design a precise current mirror with the given GaN technology, several effects such as g_m- and V_{th}-mismatch have to be considered (see Eq. 3.8). Adding a resistor as source degeneration is one option to compensate for the previously characterized g_m mismatch on

Fig. 3.16 Schematics of **a** basic current mirror and **b** current mirror with resistive source degeneration

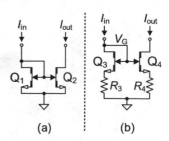

(a)   (b)

circuit level. With that, the basic current mirror in Fig. 3.16a extends to the schematic in Fig. 3.16b.

Based on the small-signal model of GaN transistors and Kirchhoff's laws, the equation for I_{out} as given in Eq. 3.8 extends to the relation in Eq. 3.9.

$$V_G = V_{GS3} + I_{in} \cdot R_3 = V_{th3} + \frac{I_{in}}{g_{m3}} + I_{in} \cdot R_3$$

$$I_{in} = \frac{V_G - V_{th3}}{R_3 + \frac{1}{g_{m3}}} = \frac{(V_G - V_{th3}) \cdot g_{m3}}{R_3 \cdot g_{m3} + 1}$$

$$I_{out} = \frac{(V_G - V_{th4}) \cdot g_{m4}}{R_4 \cdot g_{m4} + 1}$$

$$\frac{I_{out}}{I_{in}} = \frac{V_G - V_{th4}}{V_G - V_{th3}} \cdot \frac{g_{m4} \cdot (g_{m3} \cdot R_3 + 1)}{g_{m3} \cdot (g_{m4} \cdot R_4 + 1)} \approx \frac{V_G - V_{th4}}{V_G - V_{th3}} \cdot \frac{R_3}{R_4} \quad \text{for} \quad g_m \cdot R \gg 1 \quad (3.9)$$

Since $g_m \cdot R$ is typically far greater than one, the simplified expression is a good approximation. By adding a source degeneration resistor, the influence of g_m and thereby also the effects of any g_m mismatch can be compensated. The current mirror ratio is then defined by the ratio of R_3 and R_4. Since the matching of resistors (see Fig. 3.12b) is superior to the matching of g_m (Fig. 3.15b), the overall accuracy of the current mirror improves considerably. However, the influence of threshold voltage mismatch cannot be reduced by the source degeneration presented in Fig. 3.16b. This can be validated by the histograms of measured current mismatch of both current mirror implementations, depicted in Fig. 3.17.

The output current variation for the basic current mirror (Fig. 3.17a) is larger than for the current mirror with source degeneration (Fig. 3.17b). However, there is still a considerably high current mismatch despite the additional circuitry, which is most likely related to the V_{th} mismatch of adjacent GaN transistors. As derived in Eq. 3.9, the threshold voltage has a direct influence on the current mirror ratio. With the considerable high mismatch of the threshold voltages for GaN transistors (see Fig. 3.14a) the mirror ratio shows still a high variance even when the influence of g_m mismatch is fully compensated.

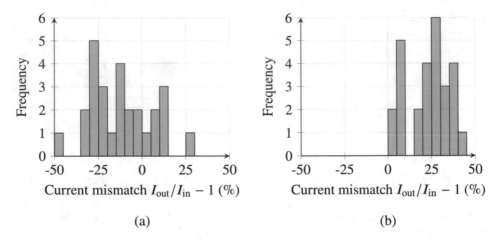

Fig. 3.17 Histogram of measured relative current mismatch for **a** basic current mirror and **b** source-degenerated current mirror

Mismatch Induced Offset of Differential Amplifiers

Another basic circuit requiring matching of both, g_m and V_{th}, is the differential pair depicted in Fig. 3.11c. It is composed of transistors Q_3 and Q_4 and a bias current source. In the differential pair, the bias current I_{bias} is distributed on the drain currents I_{d3} and I_{d4} of the two transistors depending on the input voltage difference $V_{g3} - V_{g4}$. For a circuit with ideal transistors, the current I_{bias} is distributed equally on both transistors when $V_{g3} = V_{g4}$. This is illustrated in Fig. 3.18a where the crossing of the two drain currents is located exactly at $V_{g3} = V_{g4}$. For real transistors, mismatch of g_m and V_{th} causes a shift of this current equilibrium point. The voltage difference $V_{g3} - V_{g4}$ at $I_{d3} = I_{d4} = I_{bias}/2$ is generally referred to as input-referred offset voltage V_{offset}.

The relation between V_{offset} and the transistor properties g_m and V_{th} is given by Eq. 3.10 as presented in [15]. Similar to the current mirror, the differential pair requires matching of both, g_m and V_{th} for the adjacent transistors in order to achieve low offset V_{offset}. The offset values of differently sized stand-alone differential pairs are characterized in order to assess the achievable accuracy of differential amplifiers related to the previously presented device parameter mismatch. Additionally, design techniques to compensate for the mismatch of both, V_{th} and g_m on a higher circuit level are presented.

$$V_{offset} = \left(V_{th3} + \frac{2 \cdot I_{d3}}{g_{m3}} \right) - \left(V_{th4} + \frac{2 \cdot I_{d4}}{g_{m4}} \right) \tag{3.10}$$

Figure 3.18b shows the exemplary DC characteristic for a differential pair with $V_{th3} < V_{th4}$. Due to the lower threshold voltage of Q_3, I_{d3} rises at a lower gate voltage V_{g3} compared with the ideal case in Fig. 3.18a. This leads to a negative offset voltage as predicted in Eq. 3.10. The DC characteristic of a differential pair with g_m mismatch is depicted in

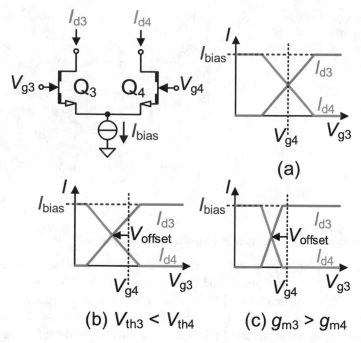

Fig. 3.18 Exemplary DC characteristic of a differential pair **a** ideal case **b** with V_{th} mismatch **c** with g_m mismatch

Fig. 3.18c. I_{d3} rises at the same gate voltage V_{g3} compared to the plot in Fig. 3.18a. However, caused by the g_m mismatch $g_{m3} > g_{m4}$, the slope of the current $I_{d3}\left(V_{g3}\right)$ is steeper. Since the bias current source forces a constant $I_{bias} = I_{d3} + I_{d4}$, this leads to a symmetrical decline of I_{d4} and a crossing of the currents for $V_{g3} < V_{g4}$. The resulting negative V_{offset} is also predicted by Eq. 3.10. Vice versa, a positive offset voltage could occur for $V_{th3} > V_{th4}$ or for $g_{m3} < g_{m4}$. A combination of both, independent V_{th} and g_m mismatch, can either cancel the shift of V_{offset} or lead to an even larger offset.

In order to evaluate the mismatch-induced offset for differential pairs in the utilized GaN technology, 58 similarly sized differential pairs are characterized across one wafer. Figure 3.19 displays a histogram of offset values for these differential pairs. All of them are laid out utilizing a common centroid style with equal current directions. Furthermore, dummy transistors are added to achieve the same neighborhood for all devices.

The characterized standard deviation of V_{offset} is approximately 70 mV and the total range of is between -200 mV and $+200$ mV. With offset values in this range, it is very challenging to design high-precision GaN circuits based on a differential pair such as comparators and operational amplifiers. However, with circuit techniques like auto-zeroing and chopping, the influence of the offset can be reduced to enable analog circuits with sufficient accuracy for integrated power converters in GaN as presented in Sect. 4.3.

Fig. 3.19 Histogram of the input referred offset voltage of 58 samples on one wafer

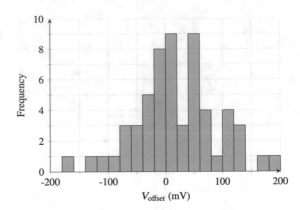

V_{offset} (mV)

Noise of Transistors, Resistors, and Fundamental Amplifiers

Besides matching properties of devices, the noise characteristic of transistors and resistors is important for the design of precise and time-invariant circuits. While noise is well investigated for d-mode GaN transistors [16, 17], literature on noise characteristics of e-mode GaN transistors is rare, especially for GaN-on-Si technology with p-GaN gate structure. However, several fluctuation effects for d-mode and e-mode GaN transistors are studied and published [18] hinting to technology-related mechanisms which may cause noise. Especially V_{th} instabilities can lead to drain current noise for GaN HEMTs. In [19], some characterization results of the threshold voltage over time show considerable fluctuations with time constants ranging from tens of microseconds to tens of seconds (see also Sect. 2.2).

In [20], the noise of a discrete 7 mΩ p-GaN gate e-mode power transistor is investigated. The spectral noise density of the drain current S_{ID} shows a $1/f$ characteristic for the examined frequency range 1 to 10, 000 Hz. Furthermore, the noise level increases with higher gate–source voltage causing higher drain current bias. Both characteristics are expected from the behavior of silicon transistors. In the BSIM4 SPICE noise model given by Eq. 3.11, I_D is in the nominator. Thus, the noise power density increases with the bias current. Additionally, the frequency f is in the denominator causing the $1/f$ characteristic.

$$S_{\text{ID}} = \frac{KF \cdot I_D{}^{AF}}{C_{\text{ox}} \cdot L_{\text{eff}}^2 \cdot f^{EF}} \tag{3.11}$$

The characterization in [20] shows spectral noise densities between $S_{\text{ID}} = 3 \times 10^{-24}$ A^2/Hz and $S_{\text{ID}} = 3 \times 10^{-21}$ A^2/Hz at 1 Hz, depending on the DC drain current level. The noise reduces by one order of magnitude per decade of the frequency. However, this is characterized for a large power transistor where a lot of averaging effects can occur. Furthermore, both the input and the output capacitances are large providing some additional filtering. Thus, the noise level of small, integrated GaN transistors is expected to be considerably higher.

In order to characterize and assess the noise performance of such integrated transistors, the drain current noise of several samples is characterized at different DC current densities

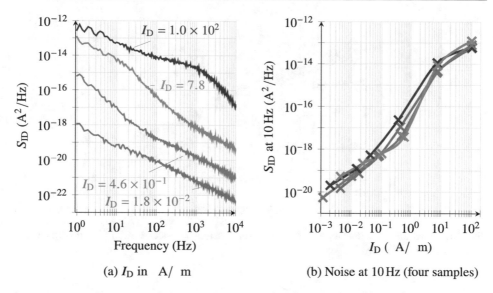

Fig. 3.20 a Spectral noise density of the drain current for different DC biases, **b** spectral noise density at $f = 10$ Hz over drain current density for four samples across one wafer

across one wafer. Figure 3.20a shows the spectral noise density of one transistor biased with different current densities. When the transistor is biased at a low current density of 1.8×10^{-2} μA/μm width, the transistor shows the lowest noise level. As expected, the noise is dominated by $1/f$-noise and, thus, the noise power density shows a decline of -10 dB per decade with increasing frequency. In contrast, the low-frequency noise increases at higher bias currents. At a current density of 7.8 μA/μm, the transistor shows an elevated noise power density at low frequencies below 40 Hz. Between 40 Hz and 600 Hz, the noise reduces with more than -10 dB per decade. At higher frequencies, the noise shows the -10 dB per decade slope again, which is expected from $1/f$-noise-dominated devices.

In order to illustrate the current dependency of the spectral noise density for the GaN transistors used in this work, the spectral noise density S_{ID} at 10 Hz is plotted over the drain current density in Fig. 3.20b. In the BSIM4 SPICE noise model given by Eq. 3.11, $S_{\mathrm{ID}} \propto I_{\mathrm{D}}^{AF}$, where AF is a correction factor. In the nominal case, $AF = 1$ [21] but it may vary in the range of 0.8–1.2. The characteristic depicted in Fig. 3.20b shows, that the GaN transistors follow the BSIM4 model at drain current densities below 0.1 μA/μm and above 10 μA/μm. However, at intermediate bias current levels, S_{ID} increases much stronger and is proportional to $I_{\mathrm{D}}^{2...3}$. Further technology research on the noise mechanisms of p-GaN gate transistors is required to understand the root causes of this noise anomaly. The influences of the transistor noise on the design of amplifier circuits in GaN is discussed further below.

Typically, a resistor is utilized as load and to bias a transistor, forming a voltage amplifier circuit in GaN (see above). Thus, in addition to the transistor noise level, also the noise of

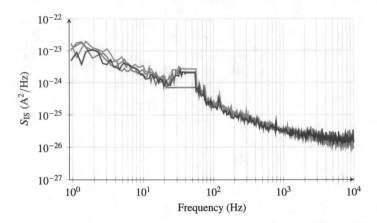

Fig. 3.21 Spectral noise density over frequency for resistors across one wafer biased with a current density of $\sim 1.5\ \mu A/\mu m$

resistors has to be characterized. Figure 3.21 shows the spectral noise density of four similar resistors across one wafer at a DC bias of $\sim 1.5\ \mu A/\mu m$.

As expected, the resistor noise is dominated by $1/f$-noise for frequencies below 1 kHz. At higher frequencies, the measurement achieves the thermal noise floor of the resistors or the measurement setup and the noise is settles at $S_{ID} =\sim \times 10^{-26}\ A^2/Hz$. The small anomaly of the plots in Fig. 3.21 for frequencies between 30 Hz and 60 Hz is related to the measurement setup, which is supplied by the 60 Hz power grid. However, the noise level of the resistor is more than seven decades lower than the noise of a transistor biased at the same current density. Thus, it is negligible for the total noise of an amplifier stage.

After the noise of resistors and transistors is characterized individually, the noise performance of a common-source amplifier composed of a transistor and a load resistor (see Fig. 3.5a) is investigated. In order to achieve a linear voltage gain above ten for the GaN amplifier, it is designed with a drain current density of $0.45\ \mu A/\mu m$. Figure 3.22a shows the spectral current noise density S_{ID} at the output of the amplifier. It matches well with the one characterized for a single transistor in Fig. 3.20a at similar drain current density of $4.6 \times 10^{-1}\ \mu A/\mu m$. Thus, the measurement confirms the assumption, that the noise of the resistor is negligible.

In order to demonstrate the effect of the spectral noise density of integrated GaN transistors on the circuit design, the common-source amplifier is also characterized by a transient measurement. Figure 3.22b shows the AC-component of the output voltage V_{out} of a common-source amplifier at DC input bias. The amplifier used for this characterization is designed with a bias resistor $R_{bias} = 200\ k\Omega$. Using the relation given in Eq. 3.3, $V_{out,AC}$ can be linearly converted to the AC-component of the drain current $I_{D,AC}$. This is represented by the right-hand side y-axis of the graph in Fig. 3.22b. The low-frequency drain current noise with an amplitude around ± 100 nA translates into voltage noise of ± 20 mV even for a comparably low resistance value of 200 kΩ.

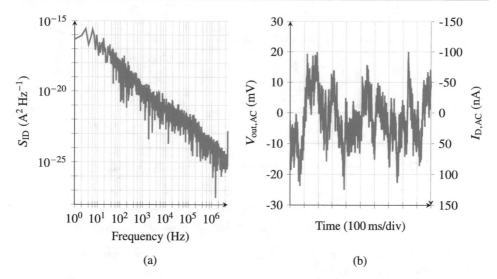

Fig. 3.22 **a** Spectral noise density of the drain current at $0.45\,\mu\text{A}/\mu\text{m}$ DC bias and **b** transient measurement of $V_{\text{out,AC}}$ and $I_{\text{D,AC}}$ of a common-source amplifier at DC input bias

The characterized noise of integrated GaN transistors significantly impairs their analog performance. It would also affect other circuits such as comparators and operational amplifiers leading to unwanted characteristics such as random offset variations. The precision of GaN circuits may improve with further research in analog design for GaN, including chopping [22, 23] and other noise canceling techniques.

Review of N-Type Only Operational Amplifiers

The presented fundamental analog circuit stages may be combined to form more powerful and complex circuits. One of the most generally useful analog circuits is the operational amplifier. It is the base for the implementation of most control loops. Considerations and concepts for operational amplifier design in a technology without complementary devices are presented in this section. The principles of design with only n-type transistors and resistors have been explored in early silicon and gallium arsenide (GaAs) designs. Many of them may be adapted for GaN technology.

Figure 3.23 depicts the schematic of an operational amplifier designed in a high-temperature GaAs technology that was proposed by [24] in 1994. It utilizes only e-mode n-type transistors and resistors, which is a major benefit for adaption in GaN technology. The circuit is designed in bipolar GaAs technology using bipolar transistors, which may be replaced by GaN HEMTs, and one diode, which can be implemented by a gate–source connected GaN transistor (see Fig. 3.1). The input stage consists of a differential amplifier based on the transistor pair Q_3 and Q_4 (comparable with Fig. 3.18) and a resistor load R_2 and R_3. The input amplifier is biased using a basic current mirror (see Fig. 3.16a) formed by Q_1 and Q_2. R_1 sets the bias current $I_{\text{bias}} = (V_{\text{DD}} - V_{\text{th}})/R_1$. Q_6 acts as source follower and

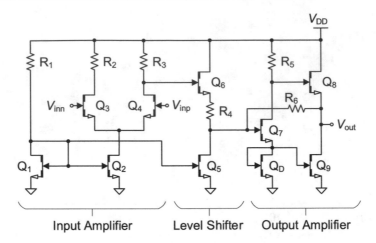

Fig. 3.23 Schematic of a resistor-transistor operational amplifier using only e-mode n-type transistors proposed for GaAs technology in [24]

forms a level shifter together with R_4 and the current source Q_5. This is required to adjust the output signal level of the input amplifier to the required input voltage level of the output amplifier. The main output amplifier is composed of Q_9 in a common-source configuration as pull-down and Q_8 as a source follower pull-up. R_6 provides a negative feedback from the output of the operational amplifier back to the pre-amplifier Q_7 and R_5. Hence, the output stage Q_8 and Q_9 achieves a low output resistance desired for operational amplifiers without spending a large quiescent current and without introducing a low-frequency pole which would be difficult to compensate within the integrated circuit [25].

The basic operational amplifier circuit in Fig. 3.23 has several limitations. The achieved open-loop gain is limited to approximately 50 dB while state-of-the-art silicon operational amplifiers typically achieve an open-loop gain well above 80 dB. Additionally, it shows a limited output voltage swing to a maximum value of $V_{DD} - V_{th}$ due to the source follower pull-up. Furthermore, the GaAs amplifier shows an input referred offset voltage of 50 mV. This is in the same range as reported for other early GaAs amplifiers with typical input-referred offset voltages of several mV [26] up to 100 mV [27]. The large offset is most likely related to the maturity level of the utilized GaAs process and the related variations of the transistor's threshold voltage.

However, several design techniques have been proposed to improve the performance of operational amplifiers without utilizing complementary devices. Figure 3.24a shows a differential amplifier implementation using only n-type transistors. This circuit is proposed in [28, 29] for silicon NMOS-only technology. In order to avoid resistor loads, diode-connected transistors Q_1 and Q_5 are used as load leading to a voltage gain of g_{m3}/g_{m1}. Therefore, the input transistors Q_3 and Q_4 are designed with a larger W/L ratio than the load transistors Q_1 and Q_5.

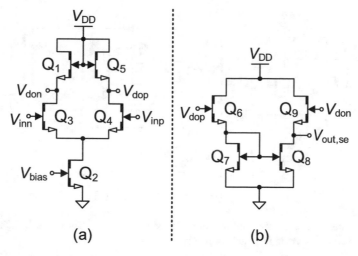

Fig. 3.24 a Schematic of a differential amplifier with n-type diode load used as input stage for an operational amplifier [28, 29]; **b** n-type only differential to single-ended converter [28, 29]

In these implementations, both outputs V_{don} and V_{dop} of the differential input amplifier are used and a dedicated differential to single-ended circuit (Fig. 3.24b) is employed to obtain a single-ended amplified voltage signal. In this conversion circuit, Q_6 provides a drain current depending on V_{dop}. This current is guided through Q_7 and mirrored to Q_8. There, it fights against the current provided by Q_9 based on V_{don} to pull down the single-ended amplified output voltage $V_{out,se}$. By utilizing these circuits, [28] achieves a gain-bandwidth product of 5 MHz and a DC gain of 51 dB. [29] reports a DC gain of 67 dB and a similar gain-bandwidth products of around 5 MHz. The average value of the input-referred offset for multiple devices is at -20 mV with a standard deviation of $\sigma = 5$ mV. A detailed analysis of all parts of an n-type only operational amplifier is provided in [25].

When the utilized process offers additional d-mode devices, a higher gain can be achieved by employing d-mode current sources as load instead of resistors or diode-connected e-mode transistors (see also Fig. 3.10). Thereby, the voltage gain of the differential input amplifier is increased, leading to a higher open-loop gain for the full amplifier. Additionally, a rail-to-rail output voltage can be achieved, when a d-mode load is utilized as pull-up. Figure 3.25a shows an output stage composed of e-mode transistors with d-mode pull-up [30]. Transistor Q_{d2} takes the role of R_6 in Fig. 3.23 to provide a negative feedback for a low output resistance with low quiescent current. The complete operational amplifier achieves an open-loop gain of 60 dB, the unity gain frequency is 3 MHz and the input-referred offset voltage is 15 mV. The relatively high offset voltage might be caused by the matching characteristics of silicon processes in the late 1970s.

Figure 3.25b depicts an input amplifier with an additional feedback loop to adjust the bias current depending on the input common-mode voltage as proposed by [31] in 1980.

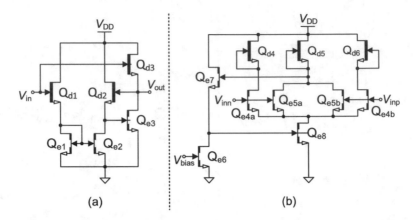

(a)   (b)

Fig. 3.25 a Schematic of a rail-to-rail output stage using d-mode pull-up [30]; **b** differential input amplifier with common mode feedback loop [31]

Table 3.1 Comparison n-type operational amplifiers

Reference	[24]	[28]	[29]	[30]	[31]
Year	1994	1976	1979	1978	1980
Technology	GaAs	Si	Si	Si	Si
Requires d-mode	No	No	No	Yes	Yes
Open-loop gain	49.5 dB	51 dB	67 dB	60 dB	70 dB
Gain-bandwidth product	n/a	5 MHz	5 MHz *	3 MHz	2.6 MHz
Offset	50 mV	$\mu = 9$ mV	$\mu = -20$ mV	< 15 mV	$\mu = -1$ mV
		$\sigma = 66$ mV	$\sigma = 5$ mV		$\sigma = 6$ mV

* estimated from plots

The biasing loop is formed by transistors Q_{e6}, Q_{e7} and Q_{e8}. When the input common mode voltage of V_{inn} and V_{inp} reduces, the common drain potential of Q_{e5a} and Q_{e5b} rises. This causes the drain current of Q_{e7} to increase, pulling the gate of Q_{e8} up. This in turn increases the bias current for the differential amplifier provided by Q_{e8} and thereby the common drain potential of Q_{e5a} and Q_{e5b} is reduced, again. Hence, the voltages at the drains of the input transistors are stabilized and the subsequent circuits can be specifically adapted to the controlled voltages. This leads to an improvement of the input-referred offset voltage to a value as low as -1 mV while achieving an open-loop gain of 70 dB and a unity gain bandwidth of 2.6 MHz (Table 3.1).

All of the presented publications on operational amplifiers in silicon and GaAs technology offer various ideas for the design of operational amplifiers without a p-type device. They provide in-depth analysis off circuit stages allowing for adaption in GaN technology with further research in the area of analog circuit design in GaN.

Review of Voltage References in GaN

GaN technologies do not offer well controlled pn-junctions (see Sect. 2.3). Hence, the established bandgap voltage reference circuit in silicon, based on the energy gap between valence and conduction band of the utilized semiconductor, can not be implemented using today's GaN technologies. However, some ideas on how a reference voltage can be generated in GaN have been published.

Wong et al. [32] presents one of the early voltage reference circuits in GaN-based on a d-mode device Q_{d1} and four Schottky barrier diodes (see Fig. 3.26a). It shows promising temperature stability of around 70 ppm/K. However, it can only generate a negative reference voltage with respect to a positive rail such as the supply voltage V_{DD}. Additionally, the generated voltage depends on absolute values of V_{th} and g_m for a single transistor. Therefore, trimming would be required to achieve a reference voltage with good absolute accuracy across several wafers and fabrication lots.

Figure 3.26b shows a circuit which generates a positive reference voltage with respect to ground. It has been presented in [33] and uses both, e-mode and d-mode devices. Q_{d2} is

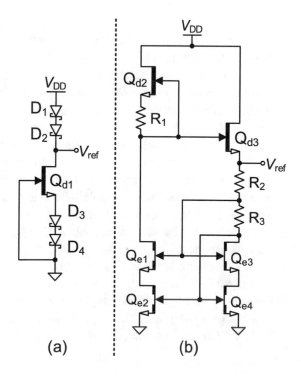

Fig. 3.26 Different voltage reference circuits suitable for GaN: **a** negative V_{ref} based on V_{th} of a d-mode transistor [32] **b** based on a d-mode current source and a temperature dependent current feedback loop [33]

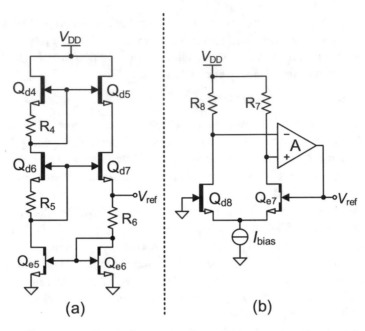

Fig. 3.27 Different voltage reference circuits suitable for GaN: **a** based on negative temperature coefficient of a d-mode current source and the positive temperature coefficient of an e-mode transistor's threshold voltage [14] **b** based on V_{th} difference of e-mode and d-mode transistors [34]

configured as current source and the current is mirrored to Q_{d3}. To generate the reference voltage, this current is guided across resistors R_2 and R_3 as well as the e-mode transistors Q_{e3} and Q_{e4}. A part of the temperature stabilization is achieved by compensating the temperature drifts of the threshold voltages of e-mode and d-mode device. Additionally, the current mirror configuration of Q_{e3} with Q_{e1} as well as Q_{e4} with Q_{e2} provides a feedback loop to control the current initially generated by Q_{d2}. The proposed circuit also requires post-fabrication trimming to achieve the desired absolute accuracy and temperature stability of 23.6 ppm/K.

A similar reference voltage circuit has been presented in 2021 [14]. It uses the CTAT characteristic of a d-mode current source and compensates this with the PTAT characteristic of an e-mode GaN transistors threshold voltage, Fig. 3.27a. Thereby, a reference voltage with low-temperature dependence is claimed; however, no data on V_{ref} over temperature or on absolute accuracy is included in the publication. Since V_{ref} is related to the threshold voltage V_{th} of a single transistor, trimming is most likely required to achieve absolute accuracy for the reference circuit across wafers and fabrication lots.

When the GaN process offers both, e-mode and d-mode devices, a ΔV_{th}-based voltage reference depicted in Fig. 3.27b might be an interesting candidate. It has been proposed for silicon technology in [34]. The reference voltage is generated based on the threshold voltage difference of e-mode and d-mode devices. Due to a similar conduction mechanism in both device types, such a circuit is likely to offer an inherently good temperature stability. However, since the fabrication of an e-mode device in GaN involves additional process steps, trimming would most likely be required to achieve absolute accuracy for the circuit. However, for silicon technology, this circuit achieves a superior temperature stability of 4.9 ppm/K making it an interesting candidate for further investigation when high-gain operational amplifiers are available in GaN technology.

If d-mode devices are not available, a reference circuit based on the transistor subthreshold slope may be considered. It typically offers a low negative temperature coefficient [35], which could be compensated by the positive temperature coefficient of GaN 2DEG resistors. [36] proposes a circuit to extract the subthreshold slope of a single device. As a major advantage, this approach avoids matching requirements by a correlated double sampling of the gate–source voltage V_{GS} of a single transistor operated at two different drain current densities. Since matching in GaN is not very accurate, yet, this technique might be a promising candidate for a voltage reference in GaN-on-Si technology. However, with the subthreshold slope as base mechanism, this circuit also relies on an absolute parameter of transistors indicating that trimming would be required in order to achieve absolute accuracy.

To summarize, today's GaN power technologies are focused on 48 V or 650 V switching performance of transistors and not on analog performance. In spite of this, analog functions can be designed and integrated into GaN technology. Regardless of the weaker analog performance in terms of single-stage gain, device mismatch and noise when comparing with silicon technology, appropriate circuit techniques such as bootstrapping, auto-zeroing, and chopping can be applied to design analog circuits with sufficient performance. This, in principle, enables GaN technology to monolithically integrate all required functionality for a power converter IC, which is further discussed in Chap. 4. However, further improvement of the fabrication process including possible extensions such as d-mode and p-type transistors as well as research on analog circuit design in GaN would further improve the integration capabilities of GaN.

3.2 Digital Integration

With the absence of p-type devices in most GaN technologies (see Sect. 2.2), state-of-the-art CMOS logic cannot be integrated in GaN. However, the available n-type transistors can be used as pull-down and resistors can be used as pull-up devices instead of p-type transistors. This resistor-transistor logic (RTL) has already been established in the late 1950s for silicon technologies [37, 38] and in the early 1980s for GaAs technologies [39], which also do not offer p-type devices. In this section, the characteristics of RTL gates in GaN are discussed and compared with silicon CMOS logic gates. Therefore, important characteristics such as edge symmetry, power consumption, area utilization, and propagation delay are investigated. Additionally, a brief review on the performance of complementary logic in academic GaN publications is provided.

Fig. 3.28 Schematics of basic
RTL logic gates

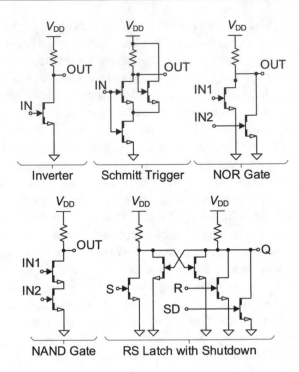

Fundamental RTL Logic Gates

In this subsection, the design of logic functions using RTL technique is presented. Several general limitations of the structure using a resistor as pull-up are discussed, especially in comparison with CMOS techniques.

Figure 3.28 shows the basic logic gates in RTL style. Similar to CMOS logic, RTL is an inverting logic where single logic gates provide an inverted output. NAND and NOR are the native combinational gates. The pull-down part composed of n-type transistors is also similar for both RTL and CMOS logic. A single pull-down transistor leads to an inverter, a series connection results in a NAND and a parallel connection of transistors forms a NOR gate. Since the pull-up of a logic gate is always one resistor and not a complementary combination of p-type transistors, effortless integration of complex logic gates with multiple NAND and NOR inputs is possible using RTL technique. Thus, a large variety of combinational logic functions can be implemented.

One limitation of RTL logic is the inherently asymmetrical behavior of the logic state transitions at the output of each gate. When the pull-down path gets turned on by a '1' at the input, the transistor has to fight against the pull-up resistor. The transistor has to be sized with an $R_{DS,on}$ much smaller than the pull-up resistance value in order to be able to pull the output low. The resistance ratio between pull-up and pull-down can be estimated depending on the input voltage level requirements of the subsequent logic gate. The threshold voltage of GaN transistors is typically around 1.3 to 2.5 V (see Sect. 2.1). Therefore, the output low level of the RTL gates should be lower than 0.5 V for all process-voltage-temperature (PVT) variations to achieve robust logic states including some margin. For nominal conditions, V_{DD} is around 5 to 6 V. Hence, the $R_{DS,on}$ of the pull-down path has to be lower than 10% of the pull-up resistance. Since the capacitive output load of a logic gate is the same for both transitions, this leads to an at least ten times faster high-to-low transition at the output of an RTL gate than achieved for the low-to-high transition.

The pull-up resistor also prevents a rail-to-rail switching of the logic gate output. When the pull-down transistor is turned off, the output goes all the way to V_{DD}. In contrast to that, the resistor is always pulling against the transistor and the output does not go all the way to ground when the pull-down transistor is turned on. Furthermore, this causes a DC quiescent current consumption as a third limitation of RTL logic gates. In contrary to CMOS logic, there is always a DC cross-current from V_{DD} to ground through the resistor and the pull-down transistor(s), when the output of the logic gate is low. A minimum-sized transistor has an $R_{DS,on}$ of up to some 10 kΩ. To maintain at least some transition similarity, the pull-up transistor cannot be sized much larger than 1 MΩ. Therefore, each logic gate draws a constant DC current from V_{DD} of $V_{DD}/R_{PU} \sim 6$ V$/1$ M$\Omega = 6$ μA when the output is low. If a larger number of logic gates is integrated, a good first-order approximation to estimate the current consumption is to assume that, all the time, around half of the logic gates are at a low output state with the associated DC current of 6 μA. Therefore, the current consumption of digital logic in GaN-based RTL technique can be estimated to be 3 μA multiplied by the total number of logic gates integrated. This puts a severe limit to the digital integration capability in GaN if high efficiency is desired, which is one major reason why GaN is at all considered for power electronics.

Another reason why digital complexity in GaN is limited is the area utilization of the required pull-up resistors. The 2DEG utilized to form resistors and transistor channels in GaN shows a specific resistance of less than 500 $\Omega/\square$ [4]. Therefore, 2000 squares are required to achieve a resistance of 1 MΩ. A small transistor is typically smaller than ten minimum squares assuming $W/L = 2$ and considering some area overhead for drain and source contacts as well as for overlap regions and contacts of the gate. Hence, the total area of an inverter is defined mostly by the area utilization of the resistor, which is more than 200 times larger than the area for a small transistor.

In silicon CMOS technology, the PMOS pull-up is typically sized with two to three times the width of the NMOS pull-down to achieve a logic gate with symmetrical transitions. This factor is used to compensate for the different mobilities of electrons and holes in silicon, leading to the same difference in specific on-resistances of complementary transistors. Starting with the same assumption that a small NMOS pull-down transistor requires an area smaller than 10 minimum squares, the complementary PMOS requires an area of less than 30 minimum squares. Hence, a small inverter in silicon CMOS technology utilizes an area of around 40 minimum squares, while a small inverter in GaN RTL technology requires more than 2000 minimum squares. This is another reason why the integration of digital functionality is less attractive in GaN technology.

The area utilization of GaN RTL logic can be reduced, if the technology offers d-mode devices. D-mode transistors can be configured easily as current source pull-up instead of using a resistor (see Fig. 3.10). The d-mode current source provides a current based on its threshold voltage V_{th} and the source resistor R_S as given by Eq. 3.6. Assuming a V_{th} of around 25% of the nominal V_{DD}, the required resistance value of R_S is also four times smaller than the value of a pull-up resistor in RTL technique for the same current. Hence, a d-mode current source requires only around 500 minimum squares instead of 2000 squares for a resistor to provide a current of 6 μA. Due to the current source characteristic, the resulting direct coupled fet logic (DCFL) also shows slightly better performance. Since it maintains a constant current as long as $V_{out} < V_{DD} - V_{th}$, the propagation delay from 0 V to $V_{DD}/2$ is $\sim 30\%$ shorter than for an RTL gate. However, the power consumption dominated by the DC cross-current as well as the non-rail-to-rail output is the same for both, RTL and DCFL technique.

Transient Performance of GaN Logic Gates

Direct measurement of transient parameters of logic gates such as propagation delay and transition times is challenging. In modern silicon technologies, logic gates achieve propagation delays well below 100 ps. Thus, an oscilloscope with a bandwidth well above 20 GHz would be required to directly measure the propagation delay with sufficient accuracy and resolution. In order to reduce the bandwidth requirements on the measurement setup, a ring oscillator formed by an odd number of same-sized inverters is employed. Hence, the average inverter delay can be calculated based on the cycle time of the ring oscillator.

The same approach is applied to characterize the propagation delay and the power consumption of RTL cells used in this work. A ring oscillator composed of 17 inverters is designed and experimentally characterized. Figure 3.29a shows a generic schematic of such an inverter-based ring oscillator. The operation frequency of the oscillator can be calculated as the inverse of the propagation delays of all inverter stages as given by Eq. 3.12.

$$f_{osc} = \frac{1}{n \cdot \left[t_{PDLH,inv} + t_{PDHL,inv} \right]}$$

$$\propto \frac{1}{n \cdot \left[(C_{gg} + C_{ds} + C_{par}) \cdot R_{PU} + (C_{gg} + C_{ds} + C_{par}) \cdot R_{DS,on} \right]} \quad (3.12)$$

For the implemented ring oscillator, n is equal to 17. The propagation delays are caused by RC time constants of each inverter cell. The capacitance is a combination of the inverter output capacitance, i.e. C_{ds} of the pull-down transistor, the gate capacitance C_{gg} of the subsequent inverter, and the parasitic capacitance C_{par} of metal interconnections. The resistance is either the value of the pull-up resistor R_{PU} or the on-resistance $R_{DS,on}$ of the pull-down transistor.

Since $R_{PU} \gg R_{DS,on}$, the propagation delay from '0' to '1' $(t_{PDLH,inv})$ is far longer than the propagation delay from '1' to '0'. This is illustrated by the timing diagram in Fig. 3.29b. For a ring oscillator to be oscillating and avoid a steady state, it is imperative that an odd

Fig. 3.29 **a** Schematic of a ring oscillator based on a RTL inverter chain and **b** exemplary timing diagram for three stages

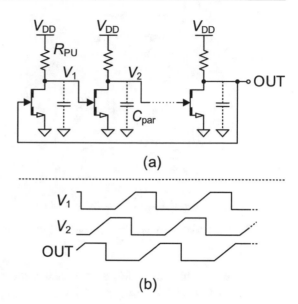

(a)

(b)

number of inverters is used. The duration of the '1'-phase at the oscillator output is defined by one more '1' to '0' transition than '0' to '1' transition in the inverter ring. Due to the very asymmetrical propagation delays of the RTL inverters, this leads to a duty cycle which is systematically smaller than 50% (see Fig. 3.29b).

Furthermore, the temperature coefficient of the GaN 2DEG resistance of typically around 12, 000 ppm/K (see Sect. 2.1) directly influences the operation frequency of the oscillator. At $-40\,°C$, the frequency is more than twice as large, while at $150\,°C$, it is less than half of the nominal frequency at room temperature.

To further simplify the experimental characterization of the ring oscillator, a T-flipflop-based frequency divider is added to the core oscillator output to slow down the frequency by a factor of 16. While the propagation delay of RTL logic gates introduces some phase shift to the divided clock signal, the frequency is accurately divided. This slower clock signal is then guided to an I/O driver and characterized with an oscilloscope.

Figure 3.30 shows the measured output voltage of the ring oscillator. The signal frequency at the pad is around 627.5 kHz, which is relatively easy to measure with a standard oscilloscope. Hence, the internal oscillation frequency f_{osc} of the ring oscillator is at ~ 10 MHz. Based on this value, the propagation delay of the inverters can be estimated by the relations given in Eq. 3.12. The sum of the propagation delays for each inverter can be calculated to $t_{\text{PDLH,inv}} + t_{\text{PDHL,inv}} \sim 5.86$ ns.

The presented way of characterizing a ring oscillator with same-sized inverters in order to calculate the average propagation delay for a single inverter is a well-established method for silicon CMOS processes [40, 41]. Table 3.2 provides an overview of characterized inverter propagation delays published for different technologies. In comparison with the GaN RTL

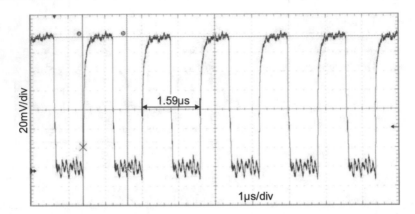

Fig. 3.30 Screenshot of an oscilloscope showing the measured output signal of the ring oscillator

inverters used in this work, CMOS inverters in a 250 nm technology show a ten times lower propagation delay and inverters in a 130 nm technology even achieve a hundred times lower propagation delay.

The value of the propagation delay depends mostly on the RC time constant formed by the pull-up resistance R_{PU} and the total load capacitance $C_{\mathrm{gg}} + C_{\mathrm{ds}} + C_{\mathrm{par}}$. R_{PU} is designed to 330 kΩ. Based on simulations, a total capacitive load of approximately 22 fF for each inverter is required to achieve the measured output frequency of 627.5 kHz. Due to the asymmetric propagation delay of RTL inverters, the ring oscillator frequency as well as the average logic gate delay depends mostly on R_{PU}. It can easily be reduced by reducing R_{PU} at the expense of more DC cross-current in each inverter. In order to assess the performance of integrated logic gates independent of the exact design and the current budget, the product of propagation delay and power consumption can be calculated as energy-delay product [40]. Thereby, the intrinsic technology capabilities can be characterized.

The utilized inverters draw a current of ~ 24 μA when the output is low. Assuming an average duty cycle of 50%, this leads to an average current consumption of ~ 12 μA from $V_{\mathrm{DD}} = 6$ V for each inverter. With $t_{\mathrm{PD}} = 5.86$ ns (see above) the average power-delay product of the inverter is $\sim 4.2 \times 10^{-13}$ W s and the energy-delay product is $\sim 2.5 \times 10^{-21}$ J s.

Besides this work, also [42] is a publication using GaN technology. However, in this publication, complementary GaN devices are utilized to form logic gates. Due to the large difference between electron and hole mobility in GaN (see Sect. 2.2), the p-type pull-up transistor is much larger. This leads to a high capacitive load at the output node and, consequently, to a relatively long propagation delay. However, with the complementary devices, lower power consumption may be achieved due to the absence of DC cross-currents of RTL gates.

Even though the presented CMOS technologies in [40, 41] are from the late 1990s and early 2000s, they are specifically engineered for best digital performance. They offer

Table 3.2 Comparison of technology properties for digital integration

	[40]	[41]	[42]	This work
Process	130 nm Si CMOS	250 nm Si CMOS	CMOS GaN	n-type GaN
Average inverter delay	<11 ps	<180 ps	51.1 ns	<2.93 ns
Power-delay product	4.5×10^{-17} W s *	6×10^{-16} W s	n/a	4.2×10^{-13} W s
Energy-delay product	5×10^{-28} J s	1.1×10^{-25} J s *	n/a	2.5×10^{-21} J s

*Calculated based on available values

devices with much smaller feature sizes and, consequently, smaller specific capacitances compared to the 650 V e-mode GaN power process utilized in this work. Additionally, standard GaN processes use silicon nitride as passivation and isolation layers [43]. With $\epsilon_r \sim 7$, silicon nitride has a much higher permittivity than silicon oxide ($\epsilon_r \sim 3.9$), which is typically used in CMOS processes. Therefore, similar structures (such as low-level routing of gate connections) show higher capacitance in GaN technology slowing down the speed of all circuits. The GaN fabrication process utilized in this work is designed to improve HV switching behavior and not high-speed low-power digital integration. Hence, Table 3.2 rather provides some general classification and not a competitive comparison of performance.

With the asymmetrical behavior, the higher power consumption and area utilization, as well as the much longer propagation delay with strong temperature dependence of RTL gates, the possibilities for digital integration in the GaN power technology used in this work are very limited. However, all commonly required logic functions can be implemented and necessary logic operations can be performed. In order to implement a power converter IC with high efficiency, the number of logic gates should be reduced to a minimum.

3.3   Mixed-Signal Integration

Mixed-signal is another category of circuits used to integrate functionality in semiconductor technologies. They generally combine both, analog and digital parts. Many state-of-the-art processes are optimized for either analog or digital circuits. Other processes, generally referred to as bipolar, CMOS and drain extended (BCD) processes, are used for most power management ICs. Besides drain-extended transistors for higher voltages, they typically offer bipolar and CMOS devices to integrate mixed-signal circuits. Today's e-mode GaN processes are typically only optimized for high-voltage switching operations and do not offer devices optimized for analog or digital operation. However, the available transistors can be used for both and, thereby, enable integration of mixed-signal circuits. In the following, design techniques without complementary devices for basic mixed-signal circuits such as comparators, analog multiplexers, and gate drivers are proposed and evaluated.

Basic Comparator

The comparator is one of the fundamental mixed-signal circuits. It compares an analog input signal, such as a voltage or a current, with a reference value and provides a digital output '1' if the input is higher than the reference and '0' when the input is lower. A single comparator can therefore be viewed as a 1-bit analog to digital converter (ADC). The design of basic comparators in silicon CMOS technology is well established and a topic in various textbooks. In contrast to that, publications on comparator design in GaN technology without a p-type device are rare.

Figure 3.31 shows some circuits, which can be employed as voltage comparators. Figure 3.31a is similar to a common source amplifier (see Fig. 3.5a) and a basic inverter (see Fig. 3.28). It can act as an inverting comparator where the reference voltage is related to the fixed value of the built-in threshold voltage V_{th} of the transistor. Depending on the sizing of resistor and transistor, the switching threshold voltage can be tuned within a small range.

In order to characterize the inverter parameters which are relevant for the application as comparator, the DC transfer curves of various sample circuits are measured. Figure 3.32 shows some DC transfer characteristic for inverter circuits with differently sized transistors in Fig. 3.32a. The results for one transistor size biased by different resistors is depicted in Fig. 3.32b (see also Sect. 3.2). The stronger the pull-down transistor and the weaker the pull-up resistor, the lower the comparator switching threshold.

Looking at it from an analog point of view, the circuit is similar to a common source amplifier and it, therefore, suffers from low voltage gain of around ten (see Fig. 3.6). For a comparator, this leads to a relatively large input voltage range of up to 1 V, where the

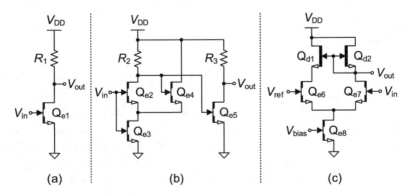

Fig. 3.31 Schematic of voltage comparator circuits: **a** inverter as comparator with fixed reference voltage $V_{ref} = V_{th}$ **b** Schmitt trigger to add a hysteresis and enhance the gain **c** variable V_{ref} comparator based on a differential amplifier [8]

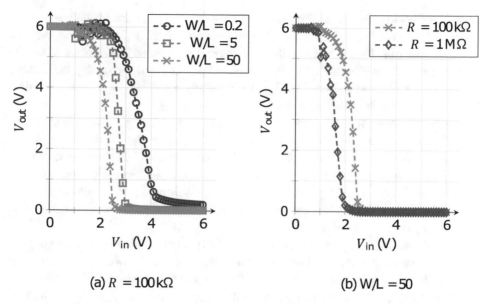

(a) $R = 100\,k\Omega$ (b) W/L $= 50$

Fig. 3.32 Measured DC transfer characteristic of an inverter **a** for different transistor sizes with a pull-up resistor of 100 kΩ; **b** for different pull-up resistors and W/L = 50

output voltage V_{out} is between 1 V and 5 V, corresponding to an undefined digital value. Additionally, the absolute value of V_{th} shows a large variation in GaN technologies of several hundred mV (see Sect. 3.1). When only low accuracy is needed and the comparator is required to decide if a voltage is below 1 V or above 4 V, this circuit is a simple and good solution. It consumes low quiescent current and is reasonably fast, since it consists only of one single stage. Furthermore, due to the low gain, no additional measures are required to avoid rapid oscillation of the comparator output as it may occur for high-gain comparators.

When higher gain is required, a positive feedback loop can be added. Figure 3.31b shows the resulting Schmitt trigger circuit. A basic inverter is added as a second stage in order to achieve a non-inverting output. The positive feedback implemented by Q_{e4} also adds a hysteresis. This is important for high-gain comparator circuits to avoid continuous toggling of the output when the input signal is close to the switching threshold.

The Schmitt trigger achieves a high gain due to the positive feedback loop implemented by Q_{e4}. When V_{in} is at ground potential, the drain of Q_{e2} and the gate of Q_{e4} is pulled to V_{DD} by R_2. Q_{e4} acts as source follower and pulls its own source together with the source of Q_{e2} to $V_{DD} - V_{th}$. Q_{e2} is thereby turned of with a negative gate–source voltage $V_{GS} = -(V_{DD} - V_{th})$. When V_{in} rises, Q_{e3} starts to turn on and reducing the source potential of Q_{e2} and Q_{e4}. Thereby, Q_{e4} with its gate still at V_{DD} experiences a higher V_{GS} and is turned on stronger. Q_{e3} fights against Q_{e4} to pull down the common source node of Q_{e2} and Q_{e4}. At some point, V_{in} is high enough that the pulling down of Q_{e3} is stronger than the pulling up of Q_{e4}. At this point, Q_{e2} starts to turn on and the gate potential of Q_{e4} reduces. This leads to the positive feedback: With lower gate potential of Q_{e4}, also the common source potential of Q_{e2} and Q_{e4} reduces which in turn increases V_{GS} of Q_{e2}. At the other transition when V_{in} goes from high to low, also the positive feedback mechanism reverses. When Q_{e2} starts to turn off, the gate of Q_{e4} is pulled towards V_{DD}. Q_{e4} turns on pulling the common source node of Q_{e2} and Q_{e4} up. Thereby, Q_{e2} is quickly turned off. A detailed analysis of such a Schmitt trigger circuit is provided in [44].

Figure 3.33 shows the DC characteristic of such a Schmitt trigger circuit implemented in GaN. The voltage gain is much larger leading to a sharp edge of the output voltage compared to the DC characteristics of the basic inverter in Fig. 3.32. For the characterization of the Schmitt trigger, V_{in} is stepped in 20 mV steps. Since no data point is located in the transition between low and high of the output voltage, this circuit has a voltage gain of at least 6 V/20 mV $= 300$ and is therefore a significant improvement over the basic inverter with a linear voltage gain of around ten.

While the Schmitt trigger shows a much higher DC gain, the switching threshold depends on the absolute value of the transistor threshold voltage V_{th} and cannot be chosen arbitrarily. However, it can be tuned in a certain range by the sizing of the transistors [45]. Furthermore, the Schmitt trigger switching threshold shows the same variations as V_{th} of several hundred millivolts (see Fig. 3.14).

Comparator circuits for arbitrary input voltages are typically based on a differential amplifier and a differential to single-ended conversion. In many cases, a latch for higher gain and hysteresis, as well as an output stage for a rail-to-rail output signal, is added. A differential voltage amplifier can be composed by adding a load to a differential pair (see Fig. 3.18). Figure 3.31c shows a comparator circuit using d-mode transistors as current source load. With the common gate node of the d-mode transistors Q_{d1} and Q_{d2} coupled to the source of Q_{d2}, the differential to single-ended conversion is directly performed in this stage [8]. In contrast to inverter and Schmitt trigger circuits, the reference voltage for this comparator

Fig. 3.33 Measured DC transfer characteristic of a GaN Schmitt trigger according to Fig. 3.31b

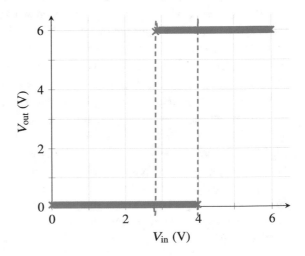

circuit is not a built-in voltage of a transistor but can be provided to the gate of one of the input transistors. However, it can only be chosen arbitrarily within certain limits to ensure proper functionality. For this circuit to operate, both e-mode transistors Q_{e6} and Q_{e7} have to be turned on to some degree. Additionally, the bias current source transistor Q_{e8} has to be operated in saturation region. For a typical gate overdrive voltage $V_{ovd} = V_{bias} - V_{th} \sim 200$ mV, the drain–source voltage of Q_{e8} needs to be greater than 200 mV. Therefore, the lower limit for the input voltage common mode at V_{ref} and V_{in} is $V_{th} + V_{ovd}$. This value is typically in the range of 1.5 to 2.5 V. The upper limit for the input common-mode range is set by the need of operation in saturation region for all devices, i.e. Q_{e6}, Q_{e7}, Q_{d1}, and Q_{d2}. Q_{d2} has the gate connected to its source. This leads to a gate overdrive $V_{ovd} = V_{GS} - V_{th,d2}$ with $V_{GS} = 0$ V and the negative threshold voltage of the depletion mode device. For operation in saturation region, the drain–source voltage of Q_{d2} needs to be greater than the gate overdrive voltage, i.e. than its threshold voltage. Similar to Q_{e8}, Q_{e7} requires a drain–source voltage greater than its overdrive, which is also typically around 200 mV. Therefore, the upper limit for the input common-mode voltage is $V_{DD} - V_{th,d2} - V_{ovd,e7} + V_{th,e7}$. This is typically some 500 mV below V_{DD}.

Besides limitations of the input common-mode voltage, also the output voltage of this circuit is limited. Due to the d-mode device used as load, the output voltage can go all the way up to V_{DD}. However, the lower boundary for V_{out} is at $V_{in} - V_{th}$ [8]. In order to achieve a rail-to-rail output signal, an additional driver stage would be required.

Another limitation is related to the matching performance of the transistors themselves. Due to the differential amplifier being used as input stage, the offset voltage of the differential pair with a standard deviation of 70 mV (see Fig. 3.19) directly translates into an offset for the comparator circuit. To achieve a higher accuracy with today's GaN process, additional circuits such as auto-zeroing or chopping are required. Inspired by further publications on comparator designs without p-type device in GaAs technology [46–48], a high-performance latched comparator with rail-to-rail input and output range as well as auto-zeroing for improved offset performance is presented in Sect. 4.3 [49, 50].

Analog Switch and Analog Multiplexer

For several circuits such as sample-and-hold stages or switched-capacitor circuits, analog voltages have to be selectively conducted or isolated. These analog voltages vary between the two supply rails V_{DD} and ground of the circuit. Thus, rail-to-rail analog switches are required to implement switched-capacitor circuits. In silicon CMOS technologies, the standard implementation of an analog switch is a transfer gate. It is composed of an n-type and a p-type transistor connected in parallel, each controlled by inverted signals at the respective gate. This way, the p-type transistor can guide voltages close to V_{DD} and the n-type transistor takes over for small voltages close to ground. The challenges for the design of transfer gates are more related to the generation of the control signals and on avoiding or compensating parasitic effects like charge injection into nodes with a high impedance.

In contrast to that, the first challenge to be considered for designing a rail-to-rail analog switch in GaN technology is the lack of suitable p-type devices (see Sect. 2.2). Utilizing

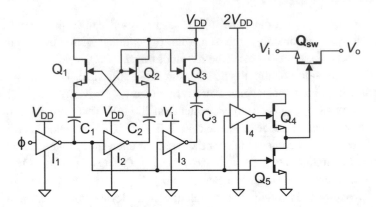

Fig. 3.34 Schematic of the bootstrapped switch without p-type device enabled by a cross-coupled charge pump

the n-type HEMT, voltages below $V_{DD} - V_{th}$ can be guided without additional efforts. For voltages closer to V_{DD}, additional circuitry is required. Possible solutions are charge pumping or bootstrapping techniques to generate a gate voltage greater than V_{DD}. This enables an n-type device to guide voltages close to V_{DD}. Since V_{DD} is typically also the maximum V_{GS}, it is advantageous to generate the gate voltage for a transistor used as analog switch with respect to its source voltage. This principle is well established for bootstrapped switches in analog-to-digital converters [51]. These bootstrapped switches are typically designed in CMOS technology employing also p-type transistors. A design of the bootstrapped switch without any p-type device is shown in Fig. 3.34 [50].

The nominal turn-on voltage V_{DD} is stored on C_3, while the switch Q_{sw} is turned off. If Q_{sw} is turned on, the input voltage V_i at the source of Q_{sw} is connected to the bottom plate of C_3 and the top plate is connected to the gate of Q_{sw} via Q_4. Therefore, V_{GS} of Q_{sw} is equal to V_{DD} and it is fully turned on, independently of the value of V_i. This bootstrapped switch can therefore be employed as a rail-to-rail analog switch. Due to leakage currents, this implementation does not support continuous operation where Q_{sw} is always turned on. If such operation is required, a second charge pump with interleaved operation can be added to recharge C_3 without turning off Q_{sw}.

The analog characteristics of the utilized GaN transistors with respect to noise and matching indicate the need for auto-zeroing and chopper circuits to achieve decent accuracy for circuits such as operational amplifiers and comparators in GaN (see Sect. 3.1). To implement auto-zeroing or chopping, an analog multiplexer is required to guide one out of two analog voltages based on a digital control input. Such an analog multiplexer can be implemented by using two bootstrapped switches (Fig. 3.34) and adding a clock gating controlled by a selection signal SEL. This is illustrated with the block diagram in Fig. 3.35.

If SEL = '0', the clock generated by an oscillator is connected to the bootstrapped switch to turn on Q_{sw1}. Thereby, V_{i1} is connected to the multiplexer output V_o. While SEL = '1',

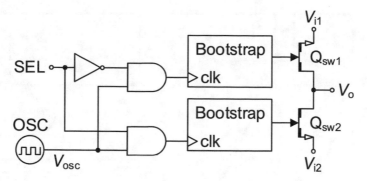

Fig. 3.35 Block diagram of an analog multiplexer based on two bootstrapped switches with clock gating

the clock is connected to the other bootstrapped switch to guide V_{i2} to the output. The oscillator is required to provide the bootstrapping clock ϕ for the active bootstrapped switch (see Fig. 3.34). The analog multiplexer does not require a dedicated oscillator but can also utilize any available clock signal, e.g. the system clock of a power converter IC.

In order to prove the concept of the proposed analog multiplexer, it is fabricated and experimentally characterized. Figure 3.36 shows measured transient waveforms of the analog multiplexer in GaN. Two different rail-to-rail analog signals are connected to the inputs of the multiplexer. V_{i1} is a sine wave at a frequency of 1 kHz and shows 6 V peak to peak voltage. V_{i2} is a saw tooth waveform with a base frequency of 10 kHz and the same 6 V peak to peak voltage.

Figure 3.36a shows the case when the selection bit SEL is set to '0' and V_{i1} is connected to the output V_o. Thereby, V_o follows the sine wave of V_{i1}, while the oscillator voltage V_{osc} is high. When V_{osc} goes low, the charge on C_3 of the bootstrapped switch is refreshed. During that time, V_o is disconnected from both V_{i1} and V_{i2} and it is slowly discharged by the DC resistance of the oscilloscope probe and input. When V_{osc} toggles to the high-voltage level again, V_o jumps nearly instantly to V_{i1} and continues to follow the input signal. For this measurement, the oscillator frequency is set to a low value of 500 Hz. It confirms the ability of the bootstrapped switch to provide sufficient gate overdrive for Q_{sw} during a time period of at least 2 ms despite the gate leakage of GaN transistors. The signal at the bottom of Fig. 3.36a shows the difference $V_{i1} - V_o$ with a noisy deviation of some tens of mV indicating a good connection between V_{i1} and V_o where most of the deviation is caused by noise in the measurement setup.

Figure 3.36b shows the waveforms for SEL = '1', where V_{i2} is connected to V_o. The 10 kHz saw tooth appears at V_o. This measurement confirms the desired selective conduction of the analog multiplexer. While the requirement for a clocked charge pump prevents time-continuous operation without additional circuitry, it is well suited for applications like auto-zeroing and chopping, which are themselves clocked operations. Furthermore, the flexibility

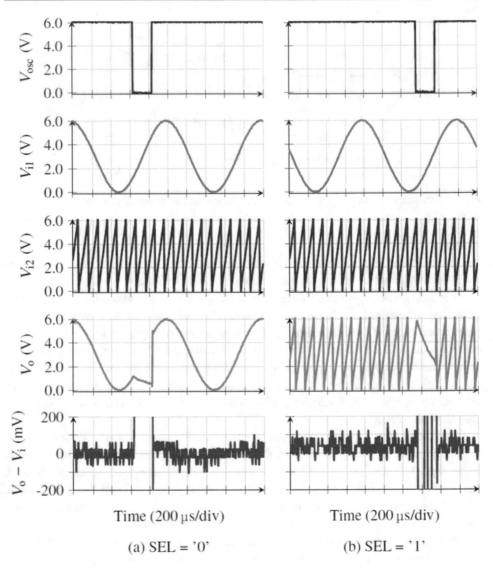

Fig. 3.36 Measured transient waveforms of the analog multiplexer **a** SEL = '0' to select V_{i1} **b** SEL = '1' to select V_{i2}

of the clock frequency to values below 500 Hz (Fig. 3.36) allows for a wide application range of the bootstrapped switch and the analog multiplexer.

Gate Driver

The combination of fast switching transients at the drain–source voltage as well as the comparably low gate charge and threshold voltage of GaN transistors puts tight limits on the maximum allowable gate loop inductance for GaN drivers (see Sect. 2.4). Thus, monolithic integration of driver and power transistor is one elegant solution to minimize gate loop parasitics and consequently enable the full switching speed of GaN power transistors (see Fig. 2.14 in Sect. 2.5). However, due to the lack of suitable p-type devices in GaN technology, integrating a gate driver faces several challenges to achieve quick and robust switching of the power transistor. The following part presents different approaches to implement an integrated gate driver and discusses the benefits and limitations with respect to integration in GaN.

Figure 3.37 shows different concepts for the implementation of gate driver circuits depending on the availability of devices. In Fig. 3.37a the state-of-the-art approach in CMOS technologies is illustrated. A NMOS transistor is used as pull-down to turn the power transistor off and a PMOS is used as pull-up to turn the power transistor on. In this approach, the dominant power consumption is caused by the transient current for charging and discharging the gate capacitance. DC current is only required to compensate gate leakage of the power transistor, which can typically be neglected. The control signal EN_B can be generated using the same voltage levels as required for the gate driver, i.e. V_{DD} and ground. For improved switching behavior, additional circuitry is typically added to achieve a break-before-make operation of M_n and M_p and to avoid cross-currents through the driver output stage. Since a p-type device is required, this concept cannot be integrated into state-of-the-art GaN technologies. On contrary, a separate gate driver IC in a CMOS technology has to be designed, which requires bond wires or PCB traces to be connected to the power transistor. This adds parasitic gate loop inductance, which is undesirable.

Figure 3.37b shows an implementation suitable for monolithic integration in GaN, since it only uses n-type transistors as pull-down and resistors as pull-up. However, a DC current always flows from V_{DD} to ground [52]. Hence, the driver output signal at the gate of the power transistor is not a full swing output but the low level is limited to some 100 mV above ground potential. This is caused by the voltage drop across the on-resistance of the pull-down transistor related to the DC cross-current. Thus, the higher power consumption and comparably slow switching speed of this RTL gate-based approach (see Sect. 3.2) leads to a reduced gate driving efficiency. A benefit of this concept is, that the generation of the control signal EN is simple and requires no additional circuitry for break-before-make operation. The signal can be generated by directly utilizing the available voltage rails V_{DD} and ground.

Figure 3.37c depicts a similar implementation where the pull-up resistors are replaced by gate–source connected d-mode transistors implementing a current source characteristic (see Fig. 3.10). Such a d-mode transistor pull-up provides a slightly better switching performance than a resistor as long as the output voltage stays one threshold voltage below V_{DD}. However,

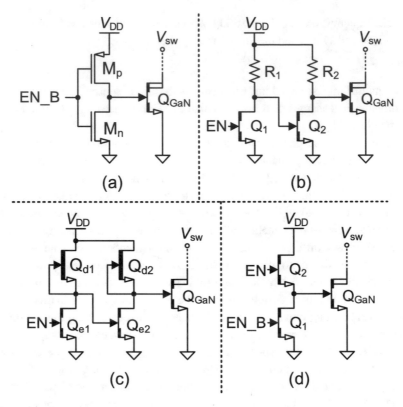

Fig. 3.37 Schematics of gate driver concepts for **a** a CMOS technology with p-type pull-up; **b** n-type only technology with resistor pull-up; **c** n-type only technology with d-mode current source pull-up; **d** n-type only technology with e-mode transistor pull-up

the pull-up cannot be turned off and causes a DC cross-current from V_{DD} to ground when the gate of the power transistor is pulled low (see also Sect. 3.2).

Figure 3.37d shows a concept where both, the pull-up and the pull-down, are implemented using n-type e-mode transistors controlled by inverted signals EN and EN_B. Thereby, a cross-current from V_{DD} to ground can be avoided and the concept provides a good trade-off between power consumption and switching speed. However, in order to achieve a rail-to-rail output where the gate voltage of the GaN power device can be pulled all the way to V_{DD}, the control signal EN for the pull-up transistor has to be generated at a voltage level greater than $V_{DD} + V_{th}$. To achieve this, different methods have been investigated in literature. [53] presents a single supply half-bridge driver including a bootstrap supply generation for the floating high-side IC. Since only the driver and the power transistor are integrated into GaN, V_{DD} can be set to a value which is a one threshold voltage higher than the nominal gate drive voltage of the high-voltage power transistor. For that approach, the EN signal is generated on the GaN die using this higher voltage level while the EN_B signal is provided externally.

Table 3.3 Overview of gate driver pull-up concepts

	PMOS	Resistor	n-type d-mode	n-type e-mode
Output range	0 V ... V_{DD}	100 mV ... V_{DD}	100 mV ... V_{DD}	0 V ... $V_{EN} - V_{th}$
Speed	High	Lowest	Low	Medium
Power consumption	Low	Highest	High	Low
Suitable for GaN integration	No	Yes	Yes	Yes

This concept manages to minimize the gate loop inductance but is not suitable for higher levels of integration in GaN technology. In [54], only the driver output stage is integrated in GaN to minimize L_p, similar to the concept shown in Fig. 3.37d. Both control signals EN and EN_B are provided externally by a pre-driver on PCB. [55] presents an integrated gate driver for a low-side power transistor, which uses an additional supply voltage rail to generate the EN signal at a higher value than the nominal gate-driver signal. The approach is suitable to support higher levels of integration, since V_{DD} is available. However, it requires the generation of an additional voltage rail for the gate driver. A similar approach is presented in [56] where a charge pump is employed to generate an additional voltage rail.

A summary and comparison of the different driver concepts illustrated in Fig. 3.37 is provided in Table 3.3.

Using an n-type e-mode transistor as pull-up is the most efficient concept for monolithic integration of a gate driver together with a power transistor in GaN technology. It offers the lowest power consumption and the highest speed out of the three concepts suitable for GaN integration. A single-supply, fully integrated, rail-to-rail gate driver according to Fig. 3.37d is presented in Sect. 4.2 [50, 57]. It utilizes bootstrapping techniques to generate the EN signal at $V_{DD} + V_{th}$ level without requiring an additional supply voltage rail.

Even though several challenges have to be addressed, the lateral e-mode GaN process enables monolithic integration of different analog, digital, and mixed-signal circuits. Additionally, it is also capable to support the integration of multiple high-voltage functions. Further investigations on monolithic integration of a power converter IC based on the considerations in this chapter are presented in Chap. 4.

References

1. Lidow, A. et al. (2012). *GaN transistors for efficient power conversion*, 1. edn. (208 pp). El Segundo, CA: Power Conversion Publications. ISBN: 9780615569253.
2. Ambacher, O. et al. (2000). Two dimensional electron gases induced by spontaneous and piezo-electric polarization in undoped and doped AlGaN/GaN heterostructures. *Journal of Applied Physics, 87.1*, 334–344. https://doi.org/10.1063/1.371866.
3. Chen, K. J. (2009). GaN smart power chip technology. In *2009 IEEE International Conference of Electron Devices and Solid-State Circuits (EDSSC)* (pp. 403–407). https://doi.org/10.1109/EDSSC.2009.5394230.
4. Hajłasz, M. et al. (2014). Sheet resistance under ohmic contacts to AlGaN/GaN heterostructures. *Applied Physics Letters, 104.24*, 242109. https://doi.org/10.1063/1.4884416.
5. Lee, J.-H. et al. (1999). DC and RF characteristics of advanced MIM capacitors for MMIC's using ultra-thin Remote-PECVD Si3N4 dielectric layers. *IEEE Microwave and Guided Wave Letters, 9.9*, 345–347. https://doi.org/10.1109/75.790469.
6. Jia, Y. et al. (2016). A robust small-signal equivalent circuit model for AlGaN/GaN HEMTs up to 110 GHz. In *2016 IEEE MTT-S International Microwave Workshop Series on Advanced Materials and Processes for RF and THz Applications (IMWS-AMP)* (pp. 1–4). https://doi.org/10.1109/IMWS-AMP.2016.7588419
7. Stockman, A., & Moens, P. (2020). ON-State gate stress induced threshold voltage instabilities in p-GaN Gate AlGaN/GaN HEMTs. In *2020 IEEE International Integrated Reliability Workshop (IIRW)* (pp. 1–4). https://doi.org/10.1109/IIRW49815.2020.9312869.
8. Liu, X., & Chen, K. J. (2011). GaN single-polarity power supply bootstrapped comparator for high-temperature electronics. *IEEE Electron Device Letters, 32.1*, 27–29. https://doi.org/10.1109/LED.2010.2088376.
9. Murmann, B. (2013). *Analysis and Design of Elementary MOS Amplifier Stages* (175 pp). National Technology and Science Press. ISBN: 9781934891179.
10. Hastings, A., & Hastings, R. A. (2006). *The art of analog layout*, 2nd edn. (648 pp). Pearson Prentice Hall. ISBN: 9780131464100.
11. Pelgrom, M. J. M., Duinmaijer, A. C. J., & Welbers, A. P. G. (1989). Matching properties of MOS transistors. *IEEE Journal of Solid-State Circuits, 24.5*, 1433–1439. https://doi.org/10.1109/JSSC.1989.572629.
12. Pelgrom, M. J. M., Tuinhout, H. P., & Vertregt, M. (1998). Transistor matching in analog CMOS applications. In *International Electron Devices Meeting 1998*. Technical Digest (Cat. No.98CH36217) (pp. 915–918). https://doi.org/10.1109/IEDM.1998.746503.
13. Chen, K. J. et al. (2017). GaN-on-Si power technology: Devices and applications. *IEEE Transactions on Electron Devices, 64.3*, 779–795.
14. Chen, H. -Y. et al. (2021). A fully integrated GaN-on-Silicon gate driver and GaN switch with Temperature-Compensated fast Turn-on technique for improving reliability. In *2021 IEEE International Solid- State Circuits Conference (ISSCC)* (pp. 460–462). https://doi.org/10.1109/ISSCC42613.2021.9365828.
15. Hellums, J. (2007). *Matching analysis and the design of low offset amplifiers*. Retrieved December 15, 2020, from https://picture.iczhiku.com/resource/eetop/WyIGydFYYoWFpmCC.pdf.
16. Silvestri, M. et al. (2013). Localization of Off-Stress-Induced damage in Al-GaN/GaN high electron mobility transistors by means of low frequency 1/f noise measurements. *Applied Physics Letters, 103.4*, 043506. https://doi.org/10.1063/1.4816424.
17. Tartarin, J. G. et al. (2013). Generation-Recombination traps in AlGaN/GaN HEMT analyzed by Time-Domain and Frequency-Domain measurements: Impact of HTRB stress on short term

and long term memory effects. In *2013 IEEE International Wireless Symposium (IWS)* (pp. 1–4). https://doi.org/10.1109/IEEEIWS.2013.6616840.

18. Meneghini, M., Meneghesso, G., & Zanoni, E., eds. (2017). *Power GaN devices* (380 pp). Switzerland: Springer International Publishing. ISBN: 9783319431970. https://doi.org/10.1007/978-3-319-43199-4.

19. Stockman, A. et al. (2019). Threshold voltage instability mechanisms in p-GaN gate AlGaN/GaN HEMTs. In *2019 31st International Symposium on Power Semiconductor Devices and ICs (ISPSD)* (pp. 287–290). https://doi.org/10.1109/ISPSD.2019.8757667.

20. Xu, X. B., et al. (2021). Analysis of trap and recovery characteristics based on Low-Frequency noise for E-Mode GaN HEMTs under electrostatic discharge stress. *IEEE Journal of the Electron Devices Society, 9,* 89–95. https://doi.org/10.1109/JEDS.2020.3040445.

21. Hung, K. K. et al. (1990). A unified model for the flicker noise in Metal-Oxide- Semiconductor Field-Effect transistors. *IEEE Transactions on Electron Devices, 37.3,* 654–665. https://doi.org/10.1109/16.47770.

22. Enz, C. C., Vittoz, E. A., & Krummenacher, F. (1987). A CMOS chopper amplifier. *IEEE Journal of Solid-State Circuits, 22.3,* 335–342. https://doi.org/10.1109/JSSC.1987.1052730.

23. Bagheri, A. et al. (2017). Low-Frequency noise and offset rejection in DCCoupled neural amplifiers: A review and Digitally-Assisted design tutorial. *IEEE Transactions on Biomedical Circuits and Systems, 11.1,* 161–176. https://doi.org/10.1109/TBCAS.2016.2539518.

24. Fricke, K. et al. (1994). AlGaAs/GaAs/AlGaAs DHBT's for High-Temperature stable circuits. *IEEE Electron Device Letters, 15.3,* 88–90. https://doi.org/10.1109/55.285393.

25. Tsividis, Y. P. (1978). Design considerations in Single-Channel MOS analog integrated Circuits—A tutorial. *IEEE Journal of Solid-State Circuits, 13.3,* 383–391. https://doi.org/10.1109/JSSC.1978.1051062.

26. Dong, Y. et al. (1995). Integrated AlGaAs/GaAs HBT high Slew-Rate and wide band operational amplifier. *IEEE Journal of Solid-State Circuits, 30.10,* 1131–1135. https://doi.org/10.1109/4.466068.

27. Katsu, S., Kazumura, M., & Kano, G. (1988). Design and fabrication of a GaAs monolithic operational amplifier. *IEEE Transactions on Electron Devices, 35.7,* 831–838. https://doi.org/10.1109/16.3333.

28. Tsividis, Y. P., & Gray, P. R. (1976). An integrated NMOS operational amplifier with internal compensation. *IEEE Journal of Solid-State Circuits, 11.6,* 748–753. https://doi.org/10.1109/JSSC.1976.1050813.

29. Young, I. A. (1979). A High-Performance All-Enhancement NMOS operational amplifier. *IEEE Journal of Solid-State Circuits, 14.6,* 1070–1077. https://doi.org/10.1109/JSSC.1979.1051317.

30. Senderowicz, D., Hodges, D. A., & Gray, P. R. (1978). High-Performance NMOS operational amplifier. *IEEE Journal of Solid-State Circuits, 13.6,* 760–766. https://doi.org/10.1109/JSSC.1978.1052047.

31. Tsividis, Y. P., Fraser, D. L., & Dziak, J. E. (1980). A Process-Insensitive High-Performance NMOS operational amplifier. *IEEE Journal of Solid- State Circuits, 15.6,* 921–928. https://doi.org/10.1109/JSSC.1980.1051498.

32. Wong, K., Chen, W., & Chen, K. J. (2010). Integrated voltage reference generator for GaN smart power chip technology. *IEEE Transactions on Electron Devices, 57.4,* 952–955. https://doi.org/10.1109/TED.2010.2041510.

33. Liao, C. et al. (2020). 3.8 A 23.6ppm/°C monolithically integrated GaN reference voltage design with temperature range from −50°C to 200°C and supply voltage range from 3.9 to 24V. In *2020 IEEE International Solid- State Circuits Conference—(ISSCC)* (pp. 72–74). https://doi.org/10.1109/ISSCC19947.2020.9062940.

34. Blauschild, R. A. et al. (1978). A new NMOS Temperature-Stable voltage reference. *IEEE Journal of Solid-State Circuits, 13.6*, 767–774. https://doi.org/10.1109/JSSC.1978.1052048.

35. Chen, J. et al. (2020). OFF-state Drain-voltage-stress-induced VTH instability in Schottky-type p-GaN gate HEMTs. *IEEE Journal of Emerging and Selected Topics in Power Electronics*, 1–1. https://doi.org/10.1109/JESTPE.2020.3010408.

36. Lin, Y., Zhang, H., & Yoshihara, T. (2013). A CMOS Low-Voltage reference based on body effect and Switched-Capacitor technique. In: *2013 International SoC Design Conference (ISOCC)* (pp. 091–094). https://doi.org/10.1109/ISOCC.2013.6863994.

37. Simkins, Q. (1958). Transistor resistor logic circuit analysis. In *1958 IEEE International Solid-State Circuits Conference*. Digest of Technical Papers (pp. 5–6). https://doi.org/10.1109/ISSCC.1958.1155618.

38. Chao, S. C. (1959). A generalized Resistor-Transistor logic circuit and some applications. In *IRE Transactions on Electronic Computers EC-8.1* (pp. 8–12). https://doi.org/10.1109/TEC.1959.5222755.

39. Ino, M., Kurumada, K., & Ohmori, M. (1981). Threshold voltage margin of Normally-Off GaAs MESFET in DCFL circuit. *IEEE Electron Device Letters, 2.6*, 144–146. https://doi.org/10.1109/EDL.1981.25375.

40. Thompson, S. et al. (2001). An enhanced 130 nm generation logic technology featuring 60 nm transistors optimized for high performance and low power at 0.7–1.4 V. In *International Electron Devices Meeting*. Technical Digest (Cat. No.01CH37224) (pp 11.6.1–11.6.4). https://doi.org/10.1109/IEDM.2001.979479.

41. Nandakumar, M. et al. (1996). A 0.25μm gate length CMOS technology for 1V low power applications—device design and power/performance considerations. In: *1996 Symposium on VLSI Technology*. Digest of Technical Papers (pp. 68–69). https://doi.org/10.1109/VLSIT.1996.507796.

42. Zheng, Z. et al. (2021). Monolithically integrated GaN ring oscillator based on High-Performance complementary logic inverters. *IEEE Electron Device Letters, 42.1*, 26–29. https://doi.org/10.1109/LED.2020.3039264.

43. Liu, X., et al. (2020). Normally-off p-GaN gated AlGaN/GaN HEMTs using plasma oxidation technique in access region. *IEEE Journal of the Electron Devices Society, 8*, 229–234. https://doi.org/10.1109/JEDS.2020.2975620.

44. Smith, M. J. S. (1988). On the circuit analysis of the schmitt trigger. *IEEE Journal of Solid-State Circuits, 23.1*, 292–294. https://doi.org/10.1109/4.293.

45. Filanovsky, I. M., & Baltes, H. (1994). CMOS schmitt trigger design. *IEEE Transactions on Circuits and Systems I: Fundamental Theory and Applications, 41.1*, 46–49. https://doi.org/10.1109/81.260219.

46. Feng, S. et al. (1991). A 4 Gs/s and 10 mV latched comparator in 0.5 μm GaAs HEMT technology. In *ESSCIRC '91: Proceedings—Seventeenth European Solid-State Circuits Conference* (pp. 109–112).

47. Feng, S., & Seitzer, D. (1992). Design on high performance GaAs latched comparator for data conversion applications. In *Proceedings 1992 IEEE International Symposium on Circuits and Systems* (Vol. 1, pp. 228–231). https://doi.org/10.1109/ISCAS.1992.229972.

48. Vold, P. J. et al. (1987). High-Performance Self-Aligned Gate AlGaAs/GaAs MODFET voltage comparator. *IEEE Electron Device Letters, 8.9*, 431–433. https://doi.org/10.1109/EDL.1987.26683.

49. Kaufmann, M. et al. (2020). 18.2 A monolithic E-Mode GaN 15W 400V offline self-supplied hysteretic buck converter with 95.6% efficiency. In *2020 IEEE International Solid-State Circuits Conference-(ISSCC)*, San Francisco, CA (pp. 288–290).

50. Kaufmann, M., & Wicht, B. (2020). A monolithic GaN-IC with integrated control loop achieving 95.6% efficiency for 400 V offline buck operation. *IEEE Journal of Solid-State Circuits, 55.12*, 3446–3454.

51. Abo, A. M., & Gray, P. R. (1999). A 1.5-V, 10-bit, 14.3-MS/s CMOS pipeline analog-to-digital converter. *IEEE Journal of Solid-State Circuits, 34.5*, 599–606. https://doi.org/10.1109/4.760369.

52. Bergogne, D. et al. (2019). Integrated GaN ICs, development and performance. In *2019 21st European Conference on Power Electronics and Applications (EPE '19 ECCE Europe)* (P.1–P.8). https://doi.org/10.23919/EPE.2019.8915151.

53. Li, X. et al. (2019). Demonstration of GaN integrated Half-Bridge with On-Chip drivers on 200-mm engineered substrates. *IEEE Electron Device Letters, 40.9*, 1499–1502. https://doi.org/10.1109/LED.2019.2929417.

54. Roberts, J., Styles, J., Chen, D. (2015). Integrated gate drivers for E-mode very high power GaN transistors. In *2015 IEEE International Workshop on Integrated Power Packaging (IWIPP)* (pp 16–19). https://doi.org/10.1109/IWIPP.2015.7295967.

55. Yamashita, Y. et al. (2018). Monolithically integrated E-mode GaN-on-SOI gate driver with power GaN-HEMT for MHz-Switching. In *2018 IEEE 6th Workshop on Wide Bandgap Power Devices and Applications (WiPDA)* (pp 231–236). https://doi.org/10.1109/WiPDA.2018.8569057.

56. Tang, G. et al. (2018). High-Speed, High-Reliability GaN power device with integrated gate driver. In: *2018 IEEE 30th International Symposium on Power Semiconductor Devices and ICs (ISPSD)* (pp. 76–79).

57. Kaufmann, M., Lueders, M., & Kaya, C. (n.d.). Gate drivers and Auto-Zero Comparators (Dallas, TX). 16/942,390.

Integration of different functions in an e-mode GaN technology has several challenges. Most critical are the lack of a suitable p-type device and a relatively large mismatch of adjacent transistors in state-of-the-art processes (see Sect. 3.1). However, the lateral structure of the HV GaN transistor allows for monolithic integration. A major benefit of e-mode GaN technology for integration is the possibility of integrating several HV functions together with the main power transistor on the same die, such as a supply voltage generator and direct sensing of the switching node voltage. Since GaN shows very promising characteristics for HV switch mode power supplies (SMPS) applications, a demonstrator system is developed to investigate the possibilities and limitations for integration in GaN with today's process technology.

In this chapter, general system considerations such as input and output specifications are presented. Based on the specifications, an appropriate operation mode and control mechanism is identified in Sect. 4.1 in order to achieve high conversion efficiency. Thereby, the requirements for the circuits composing the power converter IC are defined. Main circuit blocks are presented and characterized in Sects. 4.2–4.4 before the of the full power converter is examined in Sect. 4.5. While the demonstrator used as investigation vehicle in this work is designed for one specific application, the ideas and circuits proposed are generally valid and may also be transferred to other power converters in GaN technology.

In general, a switching power converter requires an energy storage element (L_{out}), one active switch to increase the energy in the storage element (Q_S) and one rectifier (D_{fw}) allowing for energy transport to the output, while the active switch is off. During this time, the energy in the storage element reduces. A second energy storage element (C_{out}) is typically added to achieve a DC output with only small ripple voltage or current. Such a basic and generic power converter system is depicted in Fig. 4.1. There, the rectifier is drawn as diode, but the rectifying function may also be implemented by a second switch acting as synchronous rectifier. However, such a synchronous rectifier would require additional cir-

M. P. Kaufmann and B. Wicht, *Monolithic Integration in E-Mode GaN Technology*,
Synthesis Lectures on Engineering, Science, and Technology,
https://doi.org/10.1007/978-3-031-15625-0_4

cuitry such as a floating supply voltage and a level shifter and is therefore not considered in this example for a very basic power converter.

In order to regulate the output of a power converter, the duration of the two phases, one for energizing L_{out} and the other for freewheeling, have to be controlled accordingly. Thus, a controller is required, which needs some information on the output status (v_{sns}) or signals within the power stage (i_{sns} or v_{sns}) in order to decide on when to turn the active switch Q_S on and off. The controller and the sensing circuits can be integrated together with the active switch on a single semiconductor chip, thereby achieving a high level of integration with the associated benefits in terms of power density, ease-of-use, and reliability. Additionally, a gate driver (Drv) and a flipflop (FF) to store the state of Q_S are required. All of these circuits need a LV supply voltage V_{DD}. For an easy-to-use power converter, it is desirable to generate this supply voltage in the IC itself. Hence, no additional components such as external voltage regulators or auxiliary inductor windings are required on PCB.

In Fig. 4.1, the IC is configured as a low-side buck converter with the output voltage V_{out} referred to the input voltage V_{in}. This is different from the standard implementation of a buck converter where the output V_{out} is referred to ground. However, it is well established to supply LED replacements for incandescent light bulbs. There, the whole circuit is cased within the LED light bulb. Thus, the high output potential is uncritical for safe handling of the system. Furthermore, this configuration requires only one low-side IC favoring an n-type power transistor which is the only one available in GaN. Another benefit of the low-side buck converter is that the reference potential is fixed at ground and a supply voltage can be generated from the high-voltage input without the need of bootstrapping or charge pumping. Such techniques would typically be required for a floating high-side IC which uses the switching node V_{SW} as a reference potential.

In order to investigate the possibilities and limitations for integrating a power converter IC in GaN technology, an exemplary system is designed, fabricated, and characterized. A low-side buck converter for a LED load is implemented and presented in this work as one example for a power converter system. It uses an n-type power transistor as an active switch Q_S and requires only one control loop for a constant current output. The combination of both makes it a suitable choice for integration in GaN without p-type devices and where

Fig. 4.1 Block diagram of a generic power converter IC, here configured as low-side buck converter

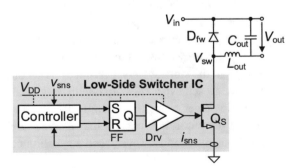

the achievable complexity is limited (see Chap. 3). However, such an IC could also be integrated in other HV semiconductor technologies. References [1, 2] are two examples of commercially available low-side buck converter ICs used for LED loads.

4.1   Demonstrator Application: LED Driver

In a GaN ICs, only a limited complexity can be reasonably integrated. This is due to the relatively large feature size of components such as transistors, resistors, and capacitors in the GaN technology as well as the comparably high quiescent current consumption, especially of digital logic gates (see Chap. 3). Therefore, a low-side buck converter for LED load with constant current output is chosen as an investigation vehicle to examine the integration capability in GaN. As a benefit for this application, no voltage control loop is required. However, it allows to demonstrate a system integration in a GaN power converter IC despite the limited complexity supported by today's GaN technologies. Additionally, several commercial silicon ICs are available for the same output specifications (e.g. [1, 2]). They provide a good base to compare the results of this work in terms of accuracy, efficiency, and power density.

The converter presented in this work shall be applicable for both, 110 and 230 V AC power grid voltage. Thus, the input voltage range is specified as V_{in} = 85 to 400 V. The desired output power is 15 W, which is the typical power consumption of a LED replacement for a 100 W incandescent light bulb to achieve similar illumination. This can be implemented by a series connection of multiple LEDs, leading to a nominal output voltage V_{out} = 50 V at a constant load current I_{out} = 300 mA. In this section, an operation mode and a suitable control method is identified to fulfill these specifications with high efficiency and appropriate accuracy. Finally, the fundamental block diagram of the implemented power converter IC is presented.

Operation Mode

Different operation modes are distinguished for switching power converters based on the nature of the inductor current. Each operation mode shows a different relationship between distinctive current values which can be measured with low effort and the average output current of the converter. Thus, the choice of the operation mode also influences the required transfer function implemented by the controller. A suitable operation mode and control method has to be identified which can be implemented in a GaN IC with its limited complexity and accuracy and, additionally, leads to a high conversion efficiency.

Figure 4.2 illustrates exemplary inductor current waveforms for the different operation modes at the same input and output specifications for a buck converter (i.e. V_{in}, V_{out}, I_{out}, L_{out}). In continuous conduction mode (CCM), the inductor current I_L ripples around the value of the output current I_{out}. The relation for I_{out} is given in Eq. 4.1.

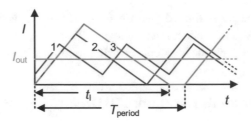

Fig. 4.2 Transient waveforms of the inductor current for (1) continuous conduction mode (CCM), (2) boundary conduction mode (BCM), and (3) discontinuous conduction mode (DCM) for the same inductance and buck converter output current I_{out}

$$I_{\text{out}} = I_{\text{peak}} - \frac{\Delta I_L}{2} = I_{\text{valley}} + \frac{\Delta I_L}{2} = \frac{I_{\text{peak}} + I_{\text{valley}}}{2} \qquad (4.1)$$

For CCM, the ripple of the inductor current is relatively small (Fig. 4.2 Waveform 1) leading to a small root mean square (RMS)-value of I_L. Therefore, the resistive losses caused by the current flow are minimal in this operation mode. However, the converter has to operate at a relatively high switching frequency to achieve CCM, which consequently increases the associated switching losses. Such switching losses are generally capacitive charging and discharging losses and therefore scale with V^2. Thus, CCM shows the most benefits and best efficiency for applications with high currents and low voltages where the current-related conduction losses dominate and the switching losses are less significant due to small voltages.

In discontinuous conduction mode (DCM) (Fig. 4.2 Waveform 3) the inductor current is greater than zero only for a part of the cycle time T_{PERIOD}. Thus, I_{valley} is zero but the relation between cycle time T_{PERIOD} and time with positive inductor current t_I affects I_{out} as shown given by Eq. 4.2.

$$I_{\text{out}} = \frac{I_{\text{peak}} + I_{\text{valley}}}{2} \cdot \frac{t_I}{T_{\text{PERIOD}}} = \frac{I_{\text{peak}}}{2} \cdot \frac{t_I}{T_{\text{PERIOD}}} \qquad \text{for } I_{\text{valley}} = 0\text{A} \qquad (4.2)$$

In this operation mode, the RMS-value of I_L is large but the switching frequency is low. Hence, DCM shows the most benefits for low-current and high-voltage applications where the switching-related losses are dominant.

The transition between CCM and DCM is called boundary conduction mode (BCM) (Fig. 4.2 Waveform 2). There the inductor current hits zero and then immediately rises again. The calculation of the output current simplifies in comparison with CCM since I_{valley} is zero and simplifies in comparison with DCM since t_I is equal to T_{PERIOD}. This is given in Eq. 4.3

$$I_{\text{out}} = \frac{I_{\text{peak}}}{2} \qquad (4.3)$$

Fig. 4.3 Simulated efficiency over output power for different operation modes at $V_{in} = 400\,V$

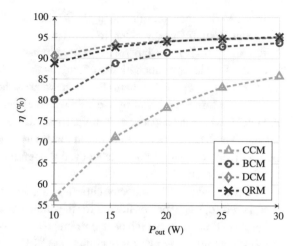

In BCM, the switching frequency and the RMS-value of I_L are located between the respective values of CCM and DCM. Therefore, BCM shows the most benefits if a trade-off between high voltages and high currents is required.

A special operation mode somewhere between BCM and DCM is quasi-resonant mode (QRM) [3]. There, a short time is allowed after I_L hits zero where the inductor current becomes slightly negative. This enables L_{out} to resonantly discharge the parasitic capacitances at the switching node V_{SW} and also to dissipate the reverse recovery charge of the freewheeling diode D_{fw} in a resonant manner. V_{SW} can drop by up to $2V_{out}$ and thereby significantly reduce the switching losses, especially at high input voltages. This comes at the cost of a slightly increased RMS-value of I_L compared to BCM.

A simplified simulation setup of a buck converter is created and the efficiency over output power for the different operation modes is plotted in Fig. 4.3 to determine the best operation mode for the given LED driver application. The simulation is performed with 400 V input voltage, a 220 µH inductor, a SiC Schottky diode as D_{fw}, a 1 Ω current sense resistor and a 1 Ω, 650 V power transistor. For the diode and the transistor, manufacturer spice models are used including their parasitic capacitances. As output load an LED model with nominal 50V at 320 mA is used.

For CCM, the switching frequency is chosen to a value leading to an inductor current ripple $\Delta I_L = 0.5 \cdot I_{out}$. Thus, it is at a high value of 2 MHz at low output power of 10 W and reduces to 670 kHz at 30 W output power. Additionally, the capacitances at the switching node are discharged across the channel resistance when the power transistor turns on. Furthermore, I-V overlap losses occur at both, the turn-on and the turn-off transition when operating in CCM due to the hard-switching characteristic of this operation mode. These losses combined with the high switching frequency in CCM lead to a low efficiency of this operation mode at the given high voltages and relatively low-current levels.

In BCM, the inductor current ripple is naturally $\Delta I_L = 2 \cdot I_{out}$, leading to a four times lower switching frequency of 500 to 168 kHz for the output power range depicted in Fig. 4.3. Additionally, no I-V overlap losses occur at turn-on since it happens at zero-current. Therefore, BCM achieves a much higher efficiency than CCM. But in BCM, the capacitances at the switching node are still hard discharged across the channel resistance when the power transistor is turned on, limiting the achievable efficiency. This is different for both DCM and QRM, where the capacitances are partially discharged in a resonant manner by the output inductor. Therefore, these operation modes achieve a considerably higher efficiency at the relatively low output power level even though the switching frequency is only slightly smaller than in BCM.

The peak value of the inductor current is one main difference between DCM and QRM and, therefore, has to be considered for the choice of a suitable operation mode. Due to the larger dead time in DCM, the peak current has to be significantly larger in order to achieve the same output current as for operation in QRM. Thus, the utilized inductor needs to be rated for a larger saturation current, which generally requires a larger physical volume of the inductor. When the converter is operated in QRM a smaller inductor can be selected, which increases the power density of the system at the cost of an only slightly reduced efficiency at very low output power (see Fig. 4.3).

To achieve high conversion efficiency for the LED buck converter with up to 400 V input voltage and 15 W nominal output power, DCM and QRM are the best choices. For higher power density, QRM is selected as the operation mode of the demonstrator system. Additionally, the complexity to integrate a controller for QRM mode is small. Only the zero crossing of the inductor current has to be sensed and no oscillator is required within the GaN IC to generate a frequency base for operation. How the output current of the converter can be controlled in QRM mode is discussed in the following part.

Control Loop

The controller has to detect the moment of zero inductor current or the first valley of the ringing at the switching node after the freewheeling diode becomes blocking in order to achieve operation in QRM. In order to regulate the average output current of a converter operating in QRM, the peak of the inductor current has to be controlled. The following part discusses a control method to achieve this behavior.

Figure 4.4 illustrates exemplary waveforms of the inductor current I_L and the switching node voltage V_{SW} in QRM. To operate in QRM, the moment to turn-on the power transistor is a short time after I_L crossed zero and V_{SW} is partially discharged. It is marked as t_{zc} in Fig. 4.4. At that point, the reverse recovery charge of the freewheeling diode has been dissipated by a resonant charge transfer. Furthermore, the charge stored at high voltages on the parasitic capacitances connected to the switching node is also dissipated in a resonant manner. Thereby, the switching losses of the converter are significantly reduced. To detect this moment, capacitive coupling can be used to sense the falling edge at V_{SW}. This method is demonstrated in [1] using a discrete capacitor. A fully integrated version of the principle is discussed in Sect. 4.3.

Fig. 4.4 Exemplary waveforms of the inductor current I_L and the switching node voltage V_{SW} in QRM

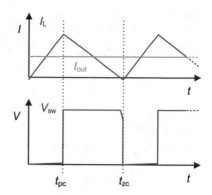

A cycle-by-cycle peak current control is utilized to control the output current I_{out} of the converter. Additionally, this control method offers inherent over current protection (OCP) for the converter. The correct moment for turning off the power transistor has to be determined to implement a peak current control. This happens when the inductor current reaches a defined peak value, which is marked as t_{pc} in Fig. 4.4. The circuit implementation of the peak current control is discussed and characterized in Sect. 4.3.

The combination of cycle-by-cycle peak current control and operation in QRM leads to a hysteretic control loop for the output current. When the inductor current crosses zero, the lower boundary of the hysteresis window is reached and the power transistor is turned on to increase the inductor current. The desired peak current represents the upper boundary of the hysteresis window where the power transistor is turned off to decrease the inductor current again. As intended by applying a hysteretic control, the inductor current as a controlled value never reaches a steady state but the low-pass filtered output current as the indirectly controlled value does. This is different from a continuous-time voltage controller typically employed in standard buck converters. There, the controller always senses the output voltage and continuously adjusts the duty cycle to meet the control condition. Hence, the duty cycle as the directly controlled value achieves a steady state. Since the power converter IC in this work is intended to be used for a LED load, only a constant current control is required and no voltage loop is needed.

Since the cycle-by-cycle peak current control is based on detecting the right moment for turning off the power transistor, any delays in the detection path lead to a systematic control error compared with respect to the simple relation given in Eq. 4.3. This is illustrated in Fig. 4.5. t_{pc} is the moment where the desired peak current is detected and the power transistor should be turned off. $t_{d,pc}$ is the delay time of the circuits performing the peak current detection and the transistor turn-off. Thus, the turn-off of the power transistor happens at $t_{pc} + t_{d,pc}$. Thereby, the real peak current and also the average output current I_{out} is increased. The current error ΔI_{peak} depends on the delay time $t_{d,pc}$, the inductance value L_{out} and the voltage across the inductor $V_{in} - V_{out}$ as given by Eq. 4.4.

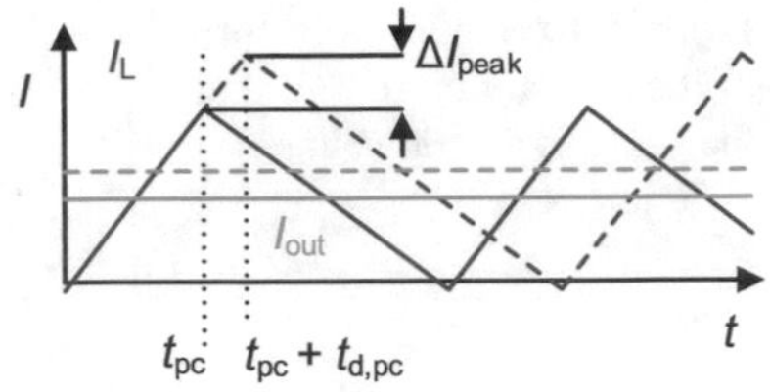

Fig. 4.5 Influence on I_{out} when the peak current turn-off is delayed

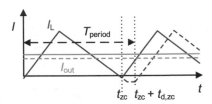

Fig. 4.6 Influence on I_{out} when the zero-current turn-on is delayed

$$\Delta I_{peak} = \frac{(V_{in} - V_{out}) \cdot t_{d,pc}}{L_{OUT}} \tag{4.4}$$

Therefore, the relation for I_{out} in Eq. 4.3 extends to the relation in Eq. 4.5.

$$I_{out} = \frac{I_{peak} + \Delta I_{peak}}{2} = \frac{I_{peak} + \dfrac{(V_{in} - V_{out}) \cdot t_{d,pc}}{L_{OUT}}}{2} \tag{4.5}$$

Another source for a control deviation of I_{out} is the delay from the moment when the inductor current crosses zero until the power transistor is turned on. This is illustrated in Fig. 4.6. Due to the delay, the operation mode shifts from QRM towards DCM. While this is intended to allow resonant dissipation of the reverse recovery charge of D_{fw} and part of the parasitic charge stored at the switching node, it also changes the relation of Eq. 4.3 to the one given in Eq. 4.6.

$$I_{out} = \frac{I_{peak}}{2} \cdot \frac{T_{period} - t_{d,zc}}{T_{period}} \tag{4.6}$$

Combining Eqs. 4.5 and 4.6, the average output current in QRM can be expressed as given by Eq. 4.7:

$$I_{out} = \frac{I_{peak} + \dfrac{(V_{in} - V_{out}) \cdot t_{d,pc}}{L_{OUT}}}{2} \cdot \frac{T_{period} - t_{d,zc}}{T_{period}} \tag{4.7}$$

This extended control equation is required since all subcircuits implementing the power converter IC cause a certain propagation delay. Most critical for the output control are the circuits for sensing the moments of zero current and peak current, the decision-making circuits as well as the gate driver turning on and off the active power switch. Based on the delay times $t_{d,pc}$ and $t_{d,zc}$, the impact of this control Eq. 4.7 changes with respect to the basic

Fig. 4.7 Block diagram of an offline buck converter IC configured as low-side buck with constant current output

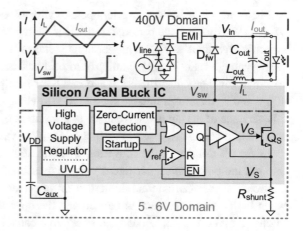

relation of Eq. 4.3. In the following sections, the delay times will be characterized based on measurement results of the circuits integrated in the GaN IC. The performance of the control is discussed in Sect. 4.5 when all parameters for Eq. 4.7 are determined.

Block Diagram

Figure 4.7 shows a high-level block diagram of the implemented GaN buck converter IC [4, 5], which is a more detailed implementation of the concept presented in Fig. 4.1. An IC represented by this block diagram could also be designed using a HV silicon technology. The IC features a 650 V power transistor together with a gate driver as power stage. A zero-current detection and a peak current comparator form the control loop for quasi-resonant operation with cycle-by-cycle peak current control. A high-voltage supply regulator is implemented for self-biased offline operation generating the supply voltage for the IC on die. All of these blocks are characterized and discussed separately in the following sections. Additionally, an UVLO is implemented to shut down the IC when the supply voltage V_{DD} is too low to ensure safe operation. An RC-based max-off-timer is included as a startup circuit for the switching operation once the supply voltage is within the desired region.

Around the IC, some discrete components are required on PCB to complete the low-side buck converter. In the 400 V domain, a diode full bridge rectifier and an electromagnetic interference EMI filter form the input stage performing a fundamental ACDC conversion. The EMI filter also includes the bulk capacitor to buffer the zero crossing of the AC power grid while the rectified line voltage ("V_{in}" in Fig. 4.7) is below V_{out}. Additionally, a freewheeling diode is placed together with the energy storage components L_{out} and C_{out} for the buck converter. Due to the implemented cycle-by-cycle peak current control and operation in QRM, the converter requires no internal time base but the switching frequency depends on external values of L_{out}, V_{out} and V_{in}. The relation is given in Eq. 4.8.

$$T = t_{on} + t_{OFF} = L_{out} \cdot \frac{I_{peak}}{V_{in} - V_{out}} + L_{OUT} \cdot \frac{I_{peak}}{V_{out}} = \frac{1}{f_{sw}} \qquad (4.8)$$

For a constant V_{out}, the converter operates with a constant off-time t_{OFF} while the on-time t_{on} and thereby the cycle period T reduces at higher input voltage V_{in}. This leads to higher switching frequencies f_{sw} for higher input voltages. The duty cycle of the converter is equal to the voltage ratio V_{out}/V_{in} similarly to all buck converters. The presented converter is specified for the typical input voltage range for grid powered voltage converters of 85 to 400 V and is, therefore, suitable for all AC grid voltages between 110 and 230 V. However, this variance of the input voltage leads to a variation of the duty cycle between 59 and 12% at an output voltage $V_{out} = 50$ V, which is a typical value for LED lighting applications.

In the 6V domain of the converter, a 1Ω resistor R_{shunt} is used to translate the current through the power transistor into a voltage measured for the cycle-by-cycle peak current control. The use of an external shunt resistor for current sensing is a well-established method when high current accuracy is desired and the operation frequency is in the hundreds of kilohertz range [6]. Additionally, one ceramic capacitor C_{aux} is used to store the supply voltage for the IC.

The GaN IC is designed with additional pads and digital multiplexers in order to allow for separate characterization of individual circuit blocks. The performance of the gate driver and the power transistor is evaluated in Sect. 4.2, the behavior of the zero-current detection and the peak current comparator is discussed in Sect. 4.3 and the on die generation of the supply voltage is covered by Sect. 4.4 before the full buck converter system is characterized in Sect. 4.5.

4.2   Characterization of the Power Stage

For minimized gate loop parasitics, the main power transistor is integrated together with its required gate driver on one die. Together, they form the power stage of the power converter IC presented in this work. In this section, both the power transistor and the gate driver are characterized individually. The results are further utilized to evaluate the implemented, delay-sensitive control loop in Sect. 4.5. Furthermore, especially the characterization results of the power transistor are important parameters for a detailed power loss analysis performed in Sect. 5.3.

An integrated 650 V GaN HEMT is employed as an active power switch for the buck converter in this work. The power switch is implemented using a 650 V e-mode GaN-on-Si technology as illustrated in Fig. 3.1 and investigated in Chap. 3. The power transistor used to energize the inductor L_{out} during the on-phase. The on-resistance RDSON of the power switch is characterized for different gate bias voltages V_{GS} by a pad-to-pad measurement on a PCB, Fig. 4.8. The measurement is performed at ambient room temperature, which is also the expected operating condition in the application as LED supply.

For this measurement, three effects can be observed:

Fig. 4.8 Measured on-resistance versus drain current of the 650 V power transistor on the GaN IC for different gate-source voltages at room temperature

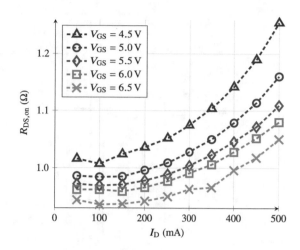

- Higher V_{GS} leads to lower $R_{DS,on}$
- Higher I_D leads to higher $R_{DS,on}$
- At lower V_{GS}, $R_{DS,on}$ increases stronger with higher I_D.

At higher V_{GS}, more electrons are accumulated below the gate forming the 2DEG channel. Therefore, the $R_{DS,on}$ reduces and the saturation current I_{SAT} of the transistor increases. At higher I_D, more and more electrons in the channel are used for the current transport. For high current densities, the current is not uniformly distributed in the channel but begins to crowd at the perimeter of the channel. This is known as the static skin effect and degenerates the linear relationship between voltage and current in a conductor. Additionally, high I_D leads to self-heating where the conduction losses in the channel increase the temperature of the transistor. Due to the large temperature coefficient of around 12000 ppm/K for the 2DEG resistance (see Sect. 2.1) this leads to higher $R_{DS,on}$ which, in turn, further increases the conduction losses. When the current is forced as it is the case in inductive switching applications, a destructive thermal run away can occur. This effect is stronger at lower V_{GS} when the initial $R_{DS,on}$ and the associated conduction losses are higher. Hence, it is important to turn-on the GaN power transistor with the highest possible voltage minimizing conduction losses and avoiding thermal run away. Thus, a rail-to-rail gate driver with a full-swing output of GND $= 0$ V and $V_{DD} = 6$ V is designed for the GaN IC.

Since the utilized GaN process does not offer p-type or d-mode devices (see Sect. 2.2), there are only two options to design a gate driver with rail-to-rail output. One option is to use resistors as pull-up leading to high DC cross currents in the driver while the output is low. The second option is to employ bootstrapped n-type transistors as pull-up. This provides higher switching speed with lower quiescent current for the gate driver (see Sect. 3.3). It is implemented according to Fig. 4.9 where Q_S is the high-voltage power transistor and the driver output-stage is formed by Q_{PD} as pull-down path and Q_{PU1} and Q_{PU2} as pull-up path [5, 7].

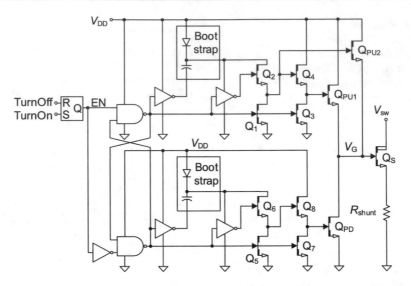

Fig. 4.9 Schematic of the single-supply rail-to-rail gate driver using bootstrapping techniques

The gate of the larger transistor Q_{PU1} is driven by V_{DD} and provides rapid turn-on by strongly pulling V_G at the gate of Q_S to $V_{DD} - V_{th}$. The gate of the smaller transistor Q_{PU2} is driven by a bootstrapped voltage greater than $V_{DD} + V_{th}$ pulling the DC level of V_G all the way to V_{DD}.

As derived in Sect. 4.1, the control method for the implemented converter is sensitive to delays (given by Eq. 4.7). Thus, the propagation delay of the implemented gate driver loaded with the gate of the power transistor is characterized. Figure 4.10a illustrates the transient waveforms of the gate driver showing a propagation delay of approximately 80 ns between the rising edge of the "Turn On" signal and the rising edge of the gate voltage V_G. Figure 4.10b shows a similar propagation delay of approximately 80 ns between the rising edge of the "Turn Off" signal and the falling edge of V_G.

The propagation delay t_{PD} itself is composed of RC delays of the different stages forming the gate driver. The output of each stage is capacitively loaded with the sum of the input capacitance of the next stage and parasitic capacitances such as metal-to-substrate capacitance of interconnections and capacitances at the drain of the transistor itself. These capacitances form an RC delay with the transistor $R_{DS,on}$ as resistance, which shows a temperature coefficient of more than 12000 ppm/k (see Sect. 2.1). Therefore, the temperature coefficient of $R_{DS,on}$ directly translates into a temperature coefficient of the propagation delay for circuits integrated in GaN (see Sect. 3.2) [8].

Figure 4.11 shows the measured propagation delay of the gate driver over temperature which varies considerably. This has to be taken into account for the design of the control loop according to Eq. 4.7. It will be discussed in detail in Sect. 4.5 when all other parameters of Eq. 4.7 are determined.

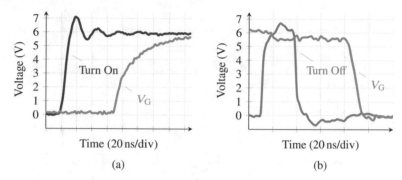

Fig. 4.10 Propagation delay of the gate driver: **a** for the turn-on and **b** for the turn-off

Fig. 4.11 Measured propagation delay of the gate driver over temperature for turn-on and turn-off

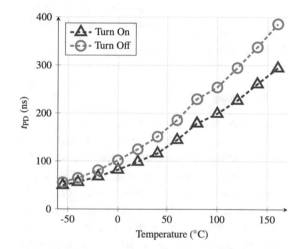

4.3   Characterization of the Analog Control Loop

For the QRM buck converter IC presented in this work, a hysteretic control is implemented according to Sect. 4.1. Hence, the moments of peak value and zero crossing of the inductor current have to be detected. The required circuits are evaluated in the following, first the implemented peak current comparator and second a zero-current detection based on direct sensing of the high-voltage switching node. One key parameter of the circuits implementing the control loop is the achievable accuracy mainly defined by the input-referred offset of the comparator. Due to the delay sensitivity of the control method (see Eq. 4.7) the propagation delay of the circuits is another critical parameter for the control loop of the converter. Thus, the offset of the comparator as well as the propagation delay of both the comparator and the zero-current detection are characterized in the following.

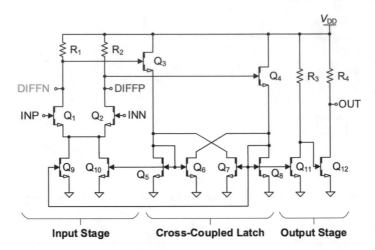

Fig. 4.12 Schematic of the core comparator

Peak Current Control

During the on-time of Q_S, the current through the transistor is equal to the inductor current I_L. It is sensed as voltage V_S across a resistor R_{shunt}, which is connected in series to the power transistor between the source terminal and ground (see Fig. 4.7). At the desired peak current set by a reference voltage V_{ref}, Q_S is required to turned off. This functionality is implemented by a peak current comparator, which senses V_S during the on-time of Q_S and compares it to a reference voltage V_{ref}. When $V_S \geq V_{ref}$, the comparator triggers a turn-off event. Thus, any input referred offset voltage of the comparator directly translates to a change of the inductor peak current and, thereby, to a control deviation of the average output current.

The core of the peak current comparator is implemented according to Fig. 4.12 [5, 7]. The main challenges for the comparator are related to the limitations of the utilized GaN process. Due to the absence of suitable p-type devices (see Sect. 2.2), it is difficult to accomplish ground-referred current sensing with a n-type input stage. Furthermore, the large threshold voltage mismatch of integrated GaN transistors (see Sect. 3.1) leads to limited accuracy of this circuit. The influence of the transistor noise is discussed further below.

To address both the low input common-mode and the limited accuracy of the core comparator with one circuit, an input-referred auto-zeroing scheme using series capacitors is implemented as illustrated in Fig. 4.13. The auto-zero loop is implemented around the differential amplifier employed as a comparator input stage. The series capacitors provide level shifting of the input voltage and, concurrently, store the offset of the input stage to compensate it when the comparator is in sensing mode.

While the power transistor Q_S is turned off, no current flows through the transistor and the peak current comparator is not required. Thus, the control signal Œ is set to '1' and the

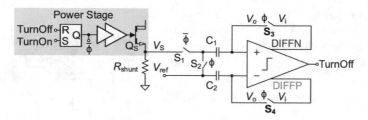

Fig. 4.13 Implementation of the input-referred auto-zero loop for offset reduction and input level shifting

comparator is in auto-zeroing mode. The switch S_1 controlled by the inverted control signal $\overline{\text{Œ}}$ is turned off to disconnect the comparator from the source of Q_S where V_S is sensed. At the same time, S_2, S_3 and S_4 (all controlled by the signal Œ) are turned on. S_2 connects the left plate of both C_1 and C_2 to the same voltage V_{ref}. Simultaneously, S_3 and S_4 put the differential amplifier into unity gain configuration and store a proper input common-mode voltage for the comparator as well as the input-referred offset on capacitors C_1 and C_2.

When Q_S is turned on, a current starts to flow through the transistor. Thus, the peak current comparator is required to sense the current value and trigger a turn-off of Q_S when the desired peak current is reached. Hence, the control signal Œ is set to "0" and the comparator exits the auto-zeroing mode. S_2, S_3 and S_4 are turned off. With a short delay, S_1 is turned on to connect the left plate of C_1 to V_S. Thereby, disturbances caused by the switching transition of the power transistor are blanked out. When S_1 is turned on, V_S is shifted by the voltage previously stored on C_1 and guided to the non-inverting input of the comparator. Similarly, V_{ref} is shifted by the voltage previously stored on C_2 and guided to the inverting input of the comparator. Thereby, a low input common-mode for V_S and V_{ref} below the device threshold voltage V_{th} can be handled. Concurrently, a possible input-referred offset of the differential amplifier is compensated by the appropriate voltages stored on C_1 and C_2.

The auto-zeroing scheme is based on the circuit presented in [9] but has to be adapted for integration in GaN technology. Especially during auto-zeroing when Œ is "1", the voltages at DIFFN and DIFFP can be close to the supply voltage V_{DD}. Therefore, the switches S_3 and S_4 cannot be implemented as single n-type transistors but have to be implemented as a bootstrapped switch (see Fig. 3.34).

Transient measurements are conducted to evaluate the input level shifting as well as the offset reducing the effect of the input-referred auto-zeroing scheme. Thus, the buck converter IC is put into a test mode according to Fig. 4.14. The drain of Q_S is left floating and V_{DD} as well as V_{ref} are provided by a DC source. While Q_S is turned off, the control signal Œ = "1" and the comparator is in auto-zero mode. To put the comparator in active sensing mode, Q_S is required to turn-on. Hence, a signal generator is used to trigger a turn-on of Q_S in a fixed pulse scheme.

After Q_S has been turned on, the source of the power transistor V_S connected to the non-inverting comparator input, is forced by a second signal generator synchronized to the

Fig. 4.14 Test setup for characterization of the peak current comparator with auto-zeroing

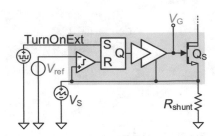

first on which controls Q_S. Different signal shapes are applied to V_S depending on which parameter of the comparator is characterized; the propagation delay or the offset. Since the signal generator with $50\,\Omega$ output impedance has a limited driving capability, the current sense resistor R_{shunt} is increased from nominal $1\,\Omega$ to a value of $1\,k\Omega$ for this test mode.

In the first step, the propagation delay of the comparator is determined. Thus, V_S is forced with a voltage step in 180% phase to the turn-on signal for Q_S. The time from the rising edge of V_S until the falling edge of V_G turning Q_S off is measured. This value includes the propagation delay of the gate driver and the RS flipflop, which has been characterized separately in Fig. 4.10 to $80\,ns$. It can be deduced from the time characterized by the measurement as shown in Fig. 4.15 to obtain the propagation delay of the comparator. Approximately $150\,ns$ pass between the crossing of V_S and V_{ref} and the falling edge of V_G. Deducing the gate driver delay of $80\,ns$, this results in a propagation delay of $70\,ns$ for the comparator at room temperature.

The other critical characteristic of the comparator is the input-referred offset, which has a direct influence on the control accuracy of the buck converter. The voltage V_S is now forced

Fig. 4.15 Propagation delay of the comparator

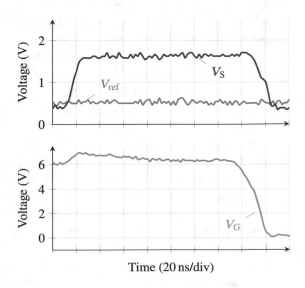

Time (20 ns/div)

with a slow ramp to characterize this offset. V_G is measured again to determine, at which voltage V_S the comparator triggers a turn-off event. The offset voltage $V_{offset} = V_S - V_{ref}$ is calculated for the time point 150 ns before the falling edge of V_G. Thereby, the total propagation delay of the comparator and gate driver is compensated. This measurement is depicted in Fig. 4.16a. The offset can be extracted from the plot to be below 5 mV. This is a major improvement compared to the offset of a differential pair characterized as typically 70 mV (one sigma) in the utilized process (see Fig. 3.19).

As discussed in Sect. 3.1, the noise of integrated GaN transistors is considerably high, especially at moderate and high DC bias of the drain current (see Fig. 3.20). Thus, a long-term measurement of the comparator offset is conducted over a time period of one second to further investigate the effects of intrinsic transistor noise on the accuracy of the comparator.

The ramp forced by the signal generator at V_S is set up with an amplitude of ± 150 mV around the reference voltage V_{ref} for this long-term measurement. Thereby, the full expected offset range for the differential pair employed as the comparator input stage is covered. When the acquire time is set to one second, the utilized oscilloscope is able to acquire and store one data point every $\Delta T = 100 \, ns$ for each signal. Based on the time resolution of the oscilloscope and the required signal amplitude at V_S, the maximum base frequency f_{VS} of the signal V_S in order to achieve a voltage resolution of $\Delta V = 1$ mV can be calculated according to Eq. 4.9.

$$f_{VS} = \frac{\Delta V}{V_{S,pp} \cdot \Delta T} = \frac{1\,mV}{300\,mV \cdot 100\,ns} \sim 33.333 \, kHz \qquad (4.9)$$

Thus, the ramp at V_S as well as the turn-on signal is provided with a base frequency of 33 kHz. With the measurement being taken for a period of one second, this leads to 33000

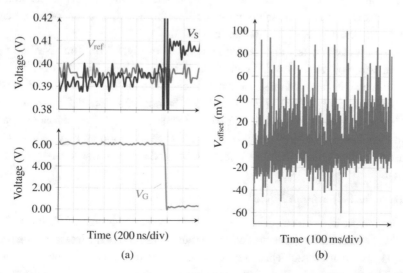

Fig. 4.16 Offset of the comparator: **a** transient plot of comparator signals, **b** offset values over one second

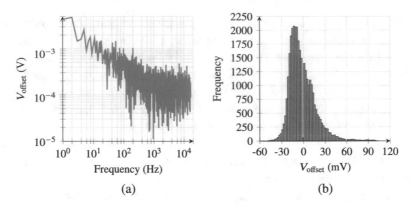

Fig. 4.17 Comparator offset analysis: **a** FFT of the offset voltage over time and **b** Histogram of 33000 measured offset voltages over one second

values of the offset. The results of this measurement are depicted in Fig. 4.16b. While most offset values are between −20 and 20 mV, the offset changes randomly over time and shows sometimes values above 80 mV. This matches with the noise of a common source amplifier characterized in Fig. 3.22b.

As a further examination of the offset, a FFT of the transient offset measurement in Fig. 4.16b is performed. The frequency spectrum is depicted in Fig. 4.17a. A $1/f$ noise is dominant in the frequency range 1 1000 Hz fitting well to the noise characterized for integrated GaN transistors in Sect. 3.1. For higher frequencies, the offset stays nearly constant indicating a thermal noise dominance. Most important is, that there is no dominant spike at a certain frequency which would indicate a systematic offset effect. Such a systematic effect could be caused by the down-sampling of an interference due to the switched-capacitor circuit forming the auto-zero loop of the comparator.

A histogram of the offset distribution is depicted in Fig. 4.17b. The distribution shows an average value of $\mu = -1.4$ mV with a standard deviation $\sigma = 18.2$ mV. The average value is within an acceptable range for the application in an LED power supply, indicating that the auto-zero loop is able to compensate systematic offset effects. However, the large standard deviation points to dynamic offset effects that can not be compensated by a sample and hold-based auto-zero loop. Design techniques such as chopping may be investigated in the future to further reduce the noise related offset and improve the accuracy of integrated GaN circuits.

Quasi Resonant Operation Mode

In order to operate in quasi-resonant mode (QRM), the power transistor Q_S has to be turned on a short time after the inductor current crosses zero at the end of the freewheeling time (t, zc in Fig. 4.4). Between the moment of zero current and the turn-on of Q_S, the inductor current reverses. The freewheeling diode, however, does not support the opposite current direction and blocks the freewheeling path. Thus, the negative inductor current rapidly discharges the

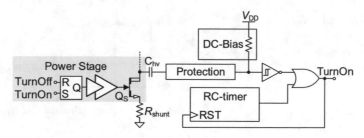

Fig. 4.18 Schematic of the implemented turn-on signal generation: Falling edge detection with protection circuits and RC-based timer for startup

parasitic capacitance at the switching node leading to a falling edge at V_{SW}. Thereby, a part of the charge stored on parasitic capacitances connected to the switching node as well as the reverse recovery charge of the freewheeling diode are dissipated in a resonant manner. This leads to smaller switching losses and consequently to higher efficiency of the implemented converter.

Thanks to the high-voltage capability of the GaN process, the switching node voltage can directly be sensed. An integrated high-voltage capacitor is used to couple the falling edge at the switching node from the 400 V domain to the low-voltage domain of the controller IC. There, the event is detected and a turn-on of the power transistor is triggered. A schematic block diagram of this circuit is shown in Fig. 4.18.

After a turn-on is triggered, Q_S rapidly pulls the switching node V_{SW} low. This falling edge at V_{SW} is much steeper and changes by a larger voltage than during resonant discharge of the switching node capacitance by the negative inductor current. Thus, a larger signal is generated by the high-voltage capacitance C_{hv} in the low-voltage domain, which may harm the input transistor gate of the following Schmitt trigger inverter. Accordingly, another large signal is generated when Q_S is turned off and V_{SW} is rapidly pushed up by the peak inductor current. Hence, a protection block consisting of series and parallel clamps is implemented between C_{hv} and the Schmitt trigger in order to prevent destructive voltages during slewing of V_{SW}.

A RC-based max-off timer (RC-Timer in Fig. 4.18, Startup in Fig. 4.7) is added as alternative way to trigger a TurnOn in order to start the clocking operation of the IC. At startup of the full converter when both V_{out} and I_L are zero, no falling edge can occur at V_{SW}. Thus, not the falling edge detection but the RC-timer triggers a turn-on of Q_S. This timer operates at a frequency much smaller than the nominal operation frequency of the buck converter to avoid interference during nominal converter operation. Hence, it is designed to trigger a turn-on after 50 μs off-time. This is approximately fifteen times larger than the off-time of the converter in nominal operation. Once the output voltage rises and the zero-current detection implemented by C_{hv} starts to work, the RC-timer is disabled. This is achieved by resetting the RC-timer at each turn-on of Q_S. Hence it does not trigger a turn-on as long as the converter is in normal operation.

Fig. 4.19 Measured signals of the zero-current detection to determine the propagation delay

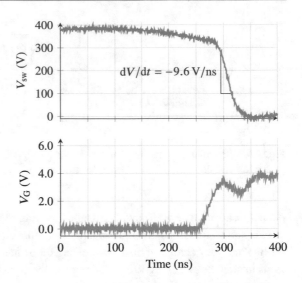

Since the implemented control loop of the converter is sensitive to delays in the turn-on path ($t_{\mathrm{d,zc}}$ in Eq. 4.7), the propagation delay of the zero-current detection circuit is characterized. It can be extracted from the transient waveforms in Fig. 4.19. When the inductor current reverses, V_{SW} shows a comparably slow discharge of starting at $\sim$125 ns in Fig. 4.19. At 250 ns, the gate voltage V_{G} of the power transistor starts to rise and is at $3\,V = V_{\mathrm{DD}}/2$ at 280 ns. At this point, the transistor is turning on and starts to pull-down V_{SW} with a much steeper slope. Between the moment of zero current and the turning on of the power transistor, $\sim$155 ns pass. This is the propagation delay of the complete turn-on path from the moment the inductor current crosses zero until Q_{S} is turned on. The turn-on path consists of the falling edge detection (Fig. 4.18) and the gate driver (Fig. 4.9). Therefore, the propagation delay of the falling edge detection can be calculated by deducing the 80 ns propagation delay of the gate driver (see Fig. 4.10) from the total propagation delay measured in Fig. 4.19. It is $\sim$75 ns long.

The propagation delay of the full turn-on path consisting of falling edge detection, RS flipflop and gate driver is $\sim$155 ns and represents $t_{\mathrm{d,zc}}$ in Eq. 4.7. The turn-off path is formed by the auto-zero comparator as well as the same RS flipflop and gate driver like used in the turn-on path. The propagation delay of the turn-off path is characterized as 150 ns ($t_{\mathrm{d,pc}}$ in Eq. 4.7). The control performance is analyzed based on these delays in Sect. 4.5 where the required values for the cycle time T as well as V_{in}, V_{out} and L_{out} are discussed.

4.4 Characterization of the Supply Concept

All circuits integrated in a power converter IC, such as the gate driver and the control loop circuits, require a supply voltage for operation. Different methods are commonly used to generate this supply voltage. Low-voltage converters may be able to use the input voltage or the generated output voltage as supply voltage. For offline converters with input voltages up to 400V and output voltages above 20V this is usually not possible since these voltages exceed the maximum allowable gate voltage of integrated transistors.

Most offline converters are isolated power converters using a transformer to generate an isolated, floating output voltage. In these applications, an auxiliary winding is typically added to the transformer generating an unregulated low-voltage supply rail in the order of 12 to 20 V. A linear voltage regulator included in the power converter IC then generates a well-controlled supply voltage such as a 5 V V_{DD} from the unregulated auxiliary voltage.

The low-side buck converter for LED loads presented in this work is a non-isolated converter since the LED light bulb is a cased system where no isolation is needed (see Sect. 4.1). Therefore, a simple power inductor with only one winding is utilized and the supply voltage for the IC has to be generated directly from the high-voltage input. Thanks to the high-voltage integration capability of the GaN process, such a high-voltage linear regulator can be integrated on the IC. This leads to two major benefits:

1. It reduces system complexity, since no additional voltage regulator or HV transistor is required on PCB
2. The regulator may track PVT variations, since it is located on the same piece of semi-conductor.

This is critical since the gate structure of GaN transistors is very sensitive to overvoltages. Higher gate voltages above 6.5 V (see Sect. 2.4) can lead to rapid aging effects or even destruction of the device [10, 11]. Therefore, a high-voltage supply regulator is integrated according to Fig. 4.20 [5, 12]. The circuit generates a supply voltage V_{DD} for the entire converter IC directly from the high-voltage input. $Q_1 - Q_6$ form a high-voltage current source to charge an external capacitor C_{aux} where the current is set by R_{lim} and the threshold voltage V_{th}. The current source turns on at each rising edge of V_{SW} due to capacitive coupling through the gate-drain capacitances of the HV-transistors.

The remaining devices form a voltage loop to limit V_{DD} to the forward voltage of D_{ref}. D_{ref} is implemented using the gate diode stack of a reference transistor (see Fig. 3.2b), which tracks PVT variations. Thereby, it provides a reference for the highest possible value of the supply voltage V_{DD} utilized for gate driving. This is important to achieve the lowest possible on-resistance for the power transistor (see Fig. 4.8) and consequently to limit the conduction losses in the transistor channel.

In order to prove the concept of the integrated supply regulator, it is characterized by means of a transient measurement of the generated supply voltage V_{DD}. The transient waveform of this measurement is depicted in Fig. 4.21. Whenever the main power transistor Q_S turns off, the current source of the supply regulator turns on and charges C_{aux} with a constant current. Thus, V_{DD} rises linearly. When V_{DD} reaches the forward voltage of D_{ref}, the charging stops and V_{DD} stays nearly constant until Q_S is turned on. Then the supply currents

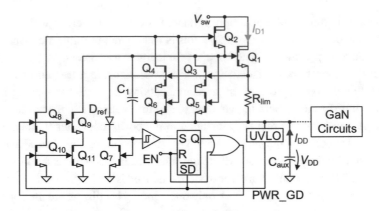

Fig. 4.20 Implementation of a high-voltage supply regulator with self turn-on

Fig. 4.21 Measurement of the supply voltage V_{DD} generated by the high-voltage supply regulator

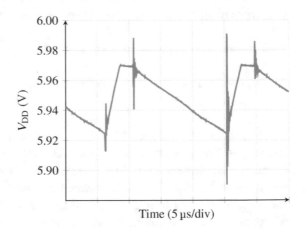

for the gate driver and the control loop discharge C_{aux} and V_{DD} declines. V_{DD} shows a ripple of 45 mV if a 470 nF ceramic capacitor is applied as C_{aux}.

Based on the capacitance value and the different slopes of V_{DD}, the currents can be estimated. The current consumption I_{DD} of the full converter IC during the on-time of Q_S can be extracted from the strongly decreasing slope. The weakly decreasing slope at the end of the freewheeling phase is related to the current consumption during the off-time of Q_S. The charging current I_{D_1} provided by the supply regulator can be calculated based on the rising slope of V_{DD}. For the calculations, the capacitance value of the employed C_{aux} is required. The utilized 10 V type [13] shows a small voltage related capacitance derating of $\sim$11% at 6 V DC bias. Together with the voltage slopes for the discharge extracted from Fig. 4.21, the supply current can be calculated based on the definition of a capacitance according to Eq. 4.10.

$$I_{DD,on} = C_{aux} \cdot \frac{1V}{1t} = 418nF \cdot \frac{41.2\,mV}{13.8\,\mu s} = 1.2\,mA \qquad (4.10)$$

During the on-time, the supply current $I_{DD,on}$ provided by C_{aux} is calculated to be 1.2 mA. In the second half of the off-time when the charging of C_{aux} is finished, (V_{DD}) shows a slower decline than during the on-time. Based on this smaller slope, $I_{DD,off}$ is calculated as 0.3 mA. In the first half of the off-time, C_{aux} is charged by the supply regulator. From the positive slope of V_{DD} the current I_{DD} can be calculated to -8.2 mA. However, during the charging, the GaN IC still draws $I_{DD,off} = 0.3$ mA. This current is directly provided by the supply regulator. Therefore, the full current I_{D1} provided by the supply regulator during charging of C_{aux} is 8.5 mA.

Depending on the input voltage V_{in} applied across the inductor L_{out} during the on-time, the low-side buck converter shows a varying operation frequency and duty cycle (see Eq. 4.8). For the wide input voltage range of 85 to 400 V the associated duty cycle varies between 59 and 12% (see Sect. 4.1). Therefore, the average supply current of the GaN IC is in the range between 400 μA and 831 μA. Since the supply current is taken from the voltage at the switching node $V_{SW} \sim V_{in}$, the power consumption of the IC varies between 70 mW at $V_{in} = 85V$ and 164 mW at $V_{in} = 400$ V. The effect of this power consumption of the IC itself on the overall converter efficiency will be discussed in Chap. 5.

4.5 System Characterization of the Full Converter

Combining the major circuits for the power stage, the control loop and the supply voltage generation presented in Sects. 4.2–4.4, a buck converter IC is implemented. It is completed by some auxiliary circuits such as a gate-drain coupled ESD active clamp as well as a blanking and filter circuit at the input of the peak current comparator (Fig. 4.22). Additionally, some digital multiplexers are added to configure the IC into a test mode to characterize different circuit blocks separately.

This section provides an overview of the system implementation of the low-side buck converter for LED lighting. The requirements for the utilized components on PCB forming the converter are discussed. Additionally, a top-level characterization of the converter is performed to evaluate the functionality and the efficiency achieved by the GaN IC. The section is concluded by a discussion of the control accuracy achieved by the implemented cycle-by-cycle peak current control based on the control relation given by Eq. 4.7

The IC is fabricated using a 650 V GaN-on-Si technology. The micrograph of the die is shown in Fig. 4.23. The total die size is 1.0 mm by 2.1 mm and includes all circuits of the full converter according to Fig. 4.22. The area of all necessary circuits for the buck converter adds up to 60% of the total die area while the test mode circuits (digital multiplexers, additional pads, ...) use 40%.

Together with the components listed in Table 4.1 the GaN IC is implemented as a low-side buck converter with constant current output for a LED load according to Fig. 4.22 in order

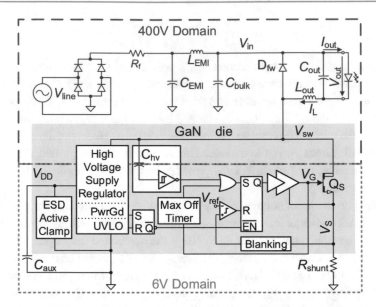

Fig. 4.22 High-level block diagram of the implemented low-side buck converter

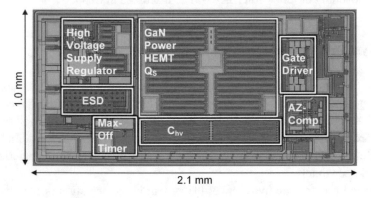

Fig. 4.23 Die micrograph of the implemented GaN buck converter IC

to evaluate the system-level performance of the converter. Based on the required voltage and current ratings for the specified application, the components are selected to achieve a converter with high efficiency and high power density.

The AC line voltage in Europe is nominal 230 V_{rms} but it is specified with 15% tolerance. Therefore, offline converters applied in Europe need to be able to handle up to 375V as peak input voltage. Thus, the input capacitors C_{EMI} and C_{bulk} are rated for 450 V allowing some additional margin to handle interferences such as ringing. When C_{EMI} is fully discharged and the converter is initially plugged in, a very high inrush current could occur harming or even destroying both the capacitor and the input rectifier. To prevent an excessively high inrush

Table 4.1 List of components for the offline buck converter

Component	Manufacturer part number	Values
Input rectifier	B10S-G	1000 V, 800 mA
R_f	5D2-10LC	5 Ω, 4A
C_{EMI}	UCS2G220MHD1TO	22 μF, 450 V
L_{EMI}	WE 7447709152	1.5mH, 400 V
C_{bulk}	UCS2G220MHD1TO	22 μF, 450 V
D_{fw}	CSD01060E-TR	600 V, 4A
C_{out}	C3225X7T2W224K200AE	2 x 220nF, 450 V
L_{out}	WE 7687714221	220 μH, 400 V
C_{aux}	GRM188R71A474KA61D	470nF, 10 V
R_{shunt}	LTR10EVHFL1R00	1 Ω, 0.5 W
LED load	SI-B8V14256HWW	48V, 320 mA

current, a negative temperature coefficient (NTC) thermistor R_f is used to limit the initial current flow. During normal operation, this resistor heats up and its resistance is reduced to a few hundred mΩ with a negligible impact on the converter efficiency. When the voltage across C_{EMI} rises quickly and C_{bulk} is completely discharged, the full 375V input voltage may be applied across L_{EMI}. Hence, a 400 V rated inductor is utilized. At startup of the IC when V_{out} is zero, the full rectified line voltage is connected across the output inductor L_{out}. Thus, it is also rated with 400 V. It has an inductance value of 220 μH to achieve a good trade-off between switching losses influenced by the switching frequency $f_{sw} \sim 1/L_{out}$ and conduction losses of the inductor DC resistance (DCR). A detailed investigation on inductor choice and the effects on losses and efficiency is provided in Sect. 5.4.

In an open-output fault scenario when no load is present to conduct the DC output current, the output voltage may rise all the way up to V_{in}. Thus, C_{out} is a 450 V ceramic capacitor and could withstand the full input voltage in such a fault scenario. The capacitance value of 440 nF is selected to achieve an output current ripple below 10% also for a reduced capacitance at $V_{out} = 50$ V DC-bias. The lowest specified input voltage is the worst case used to determine the minimum capacitance value for the input buffer capacitors. This is typically 85 V for universal voltage converters operated directly at the AC power grid. The minimum value for C_{bulk} and C_{EMI} as input buffer capacitors can be calculated according to Eq. 4.11 [14] where V_{line} is the RMS value of the AC grid voltage.

$$C_{bulk,min} + C_{EMI,min} = \frac{\frac{2 \cdot P_{in}}{f_{line,min}} \cdot \left(\frac{1}{2} - \frac{1}{2 \cdot \pi} \cdot \arccos \left(\frac{V_{in,min}}{\sqrt{2} \cdot V_{line,min}} \right) \right)}{2 \cdot V_{line,min}^2 - V_{in,min}^2} = 20.6 \, \mu F$$

$$(4.11)$$

The resulting value of $20.6\,\mu F$ is equally distributed on C_{bulk} and C_{EMI}. Considering typical tolerances of 20% and commercially available capacitors with capacitance values of the E6 series, both C_{bulk} and C_{EMI} have to be at least $15\,\mu F$ capacitors. For some increased margin, $22\,\mu F$ capacitors are selected, which can also be of a higher tolerance class.

Only two components are required outside of the converter IC in the 6 V domain. One is a 470nF capacitor C_{aux} buffering the supply voltage V_{DD} for the IC, which is generated on die by the HV supply regulator. This value is sufficient to limit the ripple of V_{DD} to less than $\pm 25\,mV$ (see Sect. 4.4). A 10V ceramic capacitor is selected for low capacitance derating at nominal $V_{DD} = 6V$ DC bias. The other component in the low-voltage domain is the current sense resistor R_{shunt}. According to the simplified output control Eq. 4.12, an error in the voltage generated by R_{shunt} has a direct effect on I_{out}. Therefore, a relatively precise resistor with 1% tolerance is selected.

$$I_{out} = \frac{I_{L,peak}}{2} = \frac{V_{ref}}{2 \cdot R_{shunt}} \tag{4.12}$$

The resistance value of $1\,\Omega$ for R_{shunt} is a compromise between conduction losses due to the current sensing as well as the accuracy and voltage level requirements for V_{ref}. The shunt resistor is rated for 0.5 W to handle the inductor peak current in the range of 600 to 700 mA.

The full converter based on the designed GaN IC is characterized for various input voltage and output current values. Figure 4.24 shows the transient waveforms during operation at $V_{in} = 400\,V$. When the gate voltage V_G is at a high level, Q_S is turned on. V_{SW} is pulled close to ground and the inductor current I_L rises. This is sensed across a $1\,\Omega$ shunt resistor as voltage V_S. At the desired peak current, the auto-zeroing comparator causes the gate driver to turn-off Q_S. The inductor current pushes the switching node up to one forward voltage of D_{fw} higher than the input voltage $(V_{in} + V_{Dfw})$ and the converter enters the freewheeling phase. Since the switching node voltage V_{SW} is pushed up by the inductor current, the (parasitic) capacitances at this node are charged with the current source characteristic of L_{out} and no capacitive hard-switching losses occur. However, there are some additional $I_L^2 \cdot R$ losses associated with the DCR of the inductor and the resistance PCB traces. A detailed analysis of the losses is provided in Sect. 5.3.

During the freewheeling phase, the inductor current reduces. When I_L reaches zero, V_{SW} stored by the switching node capacitance is still a bit higher than the input voltage V_{in} while the other terminal of the inductor is at $V_{in} - V_{out}$. This voltage across the inductor causes the current I_L to reverse and discharge the parasitic capacitances at the switching node. This falling edge is sensed and coupled to the low-voltage control by the integrated high-voltage capacitor C_{hv}. Thus, the gate driver receives a signal to turn-on Q_S again (discussed in Sect. 4.3). The reversing of the inductor current I_L slightly increases RMS losses of the inductor DCR. However, it partially discharges the capacitance at the switching node with a current source characteristic. The resonant voltage drop at V_{SW} is 60 V before the power

Fig. 4.24 Top-level waveforms of the buck operating at 400 V input voltage

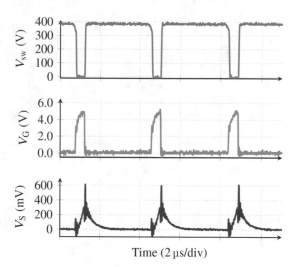

transistor is turned on (see Fig. 4.24). According to Eq. 4.13, the resonant discharge of the capacitances at the switching node C_{sw} results in a reduction of the discharging losses by 27.75% compared to a hard discharge of the full voltage.

$$\frac{P_{sw,340\,V}}{P_{sw,400\,V}} = \frac{0.5 \cdot C_{sw} \cdot 340\,V^2}{0.5 \cdot C_{sw} \cdot 400\,V^2} = \frac{340\,V^2}{400\,V^2} = 0.7225 \tag{4.13}$$

This simplified estimation is based on the assumption, that the capacitance of C_{sw} is independent of its bias voltage. A detailed investigation into the composition and voltage dependence of C_{sw} as well as the associated switching losses is provided in Sect. 5.2.

The efficiency of the implemented converter is characterized for various output current and input voltage values. Figure 4.25a shows the efficiency over an I_{out} range of 110 to 370 mA. The efficiency is above 95% for the full range indicating that the conduction losses are not the dominant losses of the implemented converter.

Figure 4.25b shows the efficiency over the input voltage at the nominal I_{out} value of 330 mA, which is important for the application as LED supply. At higher input voltage, three major effects increase the losses of the converter.

- During the on-time, the full input voltage is applied across the inductor. Due to a constant peak current turn-off, this reduces the on-time and thereby increases the switching frequency of the converter together with all associated losses (see Eq. 4.8).
- Capacitive recharging losses are proportional to V^2 (see Eq. 4.13). Thus, they increase over proportional at higher input voltage.

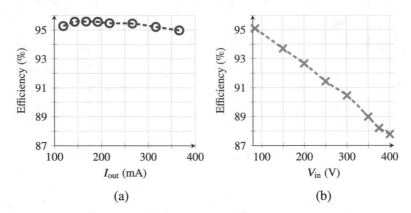

Fig. 4.25 Measured efficiency of the converter **a** over output current at $V_{in} = 85\,V$ and **b** over input voltage at $I_{out} = 330\,\text{mA}$

- The supply of the converter IC is generated from the high-voltage switching node using a linear regulator. Thus, the power consumption increases at high V_{in} due to the lower efficiency of the linear regulator (see Sect. 4.4).

However, thanks to the quasi-resonant operation mode, the switching frequency is relatively low compared to CCM (Sect. 4.1). Additionally, in QRM the capacitances are partially discharged in a resonant manner to reduce the switching losses. Therefore, high efficiency above 90% can be maintained for input voltages up to 300V. A detailed loss analysis and a comparison with a silicon implementation of a similar converter are provided in Chap. 5.

The implemented current control loop is sensitive to propagation delays relative to the cycle time of the converter (see Eq. 4.7). Thus, the achievable control accuracy based on the characterized delays and the measured cycle time of the converter is examined. Figure 4.26 shows the characterization results of different time periods as well as the operation frequency and the duty cycle of the implemented converter over its input voltage.

The converter operates in QRM with zero-current turn-on and a constant peak current control. Therefore, the inductor current ripple $\Delta I_L = I_{L,\text{peak}}$ is constant. Since the output current and thereby the output voltage across the LED load are also constant, this leads to a constant off-time t_{OFF}. This is confirmed by Fig. 4.26a, which shows a constant t_{OFF} independently of V_{in}. This is different for the on-time. According to the block diagram (Fig. 4.22) $V_{in} - V_{out}$ is applied across the inductor during t_{on}. Due to the constant peak current, this leads to a relation $t_{\text{on}} \propto 1/(V_{in} - V_{out})$. The converter cycle time T_{period} is equal to the sum of t_{on} and t_{OFF} and, therefore, follows t_{on} with an offset of $t_{\text{OFF}} = 3.35\,\mu s$. This causes the switching frequency f_{sw} to increase from 132 kHz at $V_{in} = 85\,V$ to 262 kHz at $V_{in} = 400\,V$, Fig. 4.26b. Concurrently, the duty cycle reduces from 55.9 to 11.9%. The duty cycle equals the voltage conversion ratio V_{out}/V_{in} as common for all buck converters.

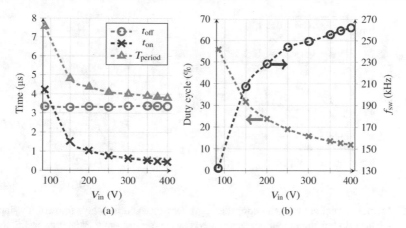

Fig. 4.26 Measured timing information over input voltage at $I_{out} = 330$ mA **a** on-time t_{on}, off-time t_{OFF} and switching period T_{period} of the converter, **b** switching frequency f_{sw} and duty cycle

Equation 4.14 (based on Eq. 4.7) gives the relation between the reference voltage V_{ref} used for the peak current detection and the output current of the buck converter operating in QRM considering the propagation delays of the turn-on and turn-off paths of the GaN IC.

$$I_{out} = \frac{\dfrac{V_{ref}}{R_{shunt}} + \dfrac{(V_{in} - V_{out}) \cdot t_{PD,off}}{L_{OUT}}}{2} \cdot \frac{T_{period} - t_{PD,on}}{T_{period}} \tag{4.14}$$

With the values for T_{period} over V_{in} from Fig. 4.26, all parameters for an analysis of the output current control based on Eq. 4.14 are available. At the nominal output current, $I_{out} = 330$ mA, the utilized LED load generates an output voltage $V_{out} = 47.5$ V. The input voltage V_{in} is specified for a range of 85 to 400 V. Over the input voltage, T_{period} varies between 7.6 μs and 3.8 μs (see Fig. 4.26a). The inductance value of the power inductor is characterized by a parameter measurement to be 198 μH. The propagation delays $t_{PD,on}$ and $t_{PD,off}$ of the control circuits are characterized as 155 and 150 ns, respectively (see Sect. 4.3).

The results of this investigation are depicted in Fig. 4.27a. It shows the resulting output current over V_{in} based on Eq. 4.14 for different values of the propagation delays $t_{PD,on} = t_{PD,off} \equiv t_{PD}$. The assumption for similar propagation delays is supported by the characterization values in Sect. 4.3. The values plotted in Fig. 4.27a represent the calculated output current for the case that a constant reference voltage V_{ref} is used for the peak current detection. This gives an indication of the achievable control accuracy based on the propagation delays of the converter IC. The plot shows an increased output current at larger input voltage. Additionally, larger propagation delays increase the control deviation at higher V_{in}. Hence, the propagation delay of the turn-off path including the peak current comparator has a stronger influence on the output current control.

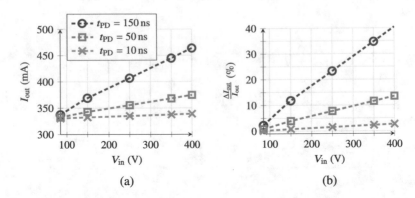

Fig. 4.27 Output current control with constant V_{ref} over propagation delays for different input voltages: **a** absolute value of the output current and **b** relative control error compared to the ideal I_{out}

Figure 4.27b shows the relative control error between the ideal output current of 330 mA and the calculated values in Fig. 4.27a. For the characterized propagation delays of $\sim$150 ns the current control with a constant V_{ref} is only sufficiently accurate (below 5%) for $V_{in} = 85$ V. To maintain a good output current accuracy for the GaN-based buck converter comparable to state-of-the-art silicon converters [1, 2] one way would be to reduce the propagation delay of the turn-on and turn-off paths by 90% to 15 ns. This becomes an even more ambitious goal since the propagation delay would have to be below that value for the full specified temperature range where the 2DEG temperature coefficient leads to a significant variation of the propagation delays by a factor of four (see Sect. 4.2). Another possible way would be to implement a compensation of V_{ref} based on either V_{in} or on a feedback loop sensing the actually achieved peak current when the power transistor turns off.

Summarizing the characterization results presented in this chapter, the integration of analog functions, high-voltage sensing, and a supply voltage regulator on one power converter IC using GaN technology is possible. Technology related challenges such as the lack of suitable complementary transistors as well as the larger device mismatch are handled on circuit and system level. The implemented GaN IC achieves high efficiency up to 95.6%. Improvements towards better matching and lower propagation delays are expected with further research in the area of circuit design and process technology in GaN. Thereby, GaN would not only show superior power efficiency and switching performance compared with silicon but might also become competitive in terms of control accuracy.

References

1. Power Integrations. (2016). Single-Stage LED driver IC with combined PFC and constant current output for buck topology. In LYT1402-1604 LYTSwitch-1 Family Datasheet.
2. STMicroelectronics. (2014). Offline LED driver with primary-sensing and high power factor up to 15 W. In HVLED815PF Datasheet.

3. Faraci, E. et al. (2016). High efficiency and power density GaN-based LED driver. In *2016 IEEE Applied Power Electronics Conference and Exposition (APEC)* (pp. 838–842).
4. Kaufmann, M. et al. (2020). 18.2 a monolithic E-mode GaN 15 W 400 V offline self-supplied hysteretic buck converter with 95.6
5. Kaufmann, M., & Wicht, B. (2020). A Monolithic GaN-IC with Integrated Control Loop Achieving 95.6.
6. Wickramasinghe, T. et al. (2019). A study on shunt resistor-based current measurements for fast switching GaN devices. In *IECON 2019—45th Annual Conference of the IEEE Industrial Electronics Society* (Vol. 1, pp. 1573–1578). https://doi.org/10.1109/IECON.2019.8927490.
7. Kaufmann, M., Lueders, M., & Kaya, C. (n.d.). Gate Drivers and Auto-Zero Comparators. 16/942,390 (Dallas, TX).
8. Kaufmann, M., Seidel, A., & Wicht, B. (2020). Long, short, monolithic—the gate loop challenge for GaN drivers: Invited paper. In *2020 IEEE Custom Integrated Circuits Conference (CICC)*, Boston, MA (pp. 1–5).
9. Pei, R. et al. (2017). A low-offset dynamic comparator with input offset-cancellation. In *2017 IEEE 12th International Conference on ASIC (ASICON)* (pp. 132–135). https://doi.org/10.1109/ASICON.2017.8252429.
10. Wu, T. et al. (2015). Forward bias gate breakdown mechanism in enhancement-mode p-GaN gate AlGaN/GaN high-electron mobility transistors. In: *IEEE Electron Device Letters*, (Vol. 36.10, pp. 1001–1003). DOIurl10.1109/LED.2015.2465137.
11. Rossetto, I. et al. (2016). Experimental demonstration of weibull distributed failure in p-type GaN high electron mobility transistors under high forward bias stress. In *2016 28th International Symposium on Power Semiconductor Devices and ICs (ISPSD)* (pp. 35–38). https://doi.org/10.1109/ISPSD.2016.7520771.
12. Lueders, M. et al. (n.d.). Enhancement Mode Startup Circuit with JFET Emulation. 16/731,847 (Dallas, TX).
13. Murata. (2015). GRM188R71A474KA61. In: *Capacitor Data Sheet.* https://datasheet.octopart.com/GRM188R71A474KA61D-Murata-datasheet-59261961.pdf.
14. Texas Instruments. (2016). UCC28880 700-V, 100-mA low quiescent current off-line converter. In UCC28880 Datasheet.

Performance Comparison of Monolithic GaN and Silicon Converters

5

In this work, a monolithically integrated buck converter IC is developed and fabricated using an e-mode GaN technology (see Chap. 4). In order to learn more on how the use of GaN can improve the performance of offline power converters, the fabricated IC is compared with a silicon buck converter IC presented in Sect. 5.1. Therefore, both the GaN and the silicon ICs are assembled together with the same external components to form similar buck converters. A model for the losses of the power converters is developed based on datasheet parameters and various measurements according to Fig. 5.1.

The gray boxes on the left provide an overview of different datasheet parameters utilized in the loss model to examine the mechanisms and distribution of power loss in the converters. These parameters include various values for capacitances and resistances as well as I-V curves of diodes. The different measurement results provided for the loss model are captured with the blue boxes on the right. They include DC measurements of voltages and currents at the input and the output as well as steady-state values during operation of the converter, such as the switching frequency f_{sw} and the duty cycle. Transient measurements of the turn-on and turn-off transitions are also included to validate the capacitance values provided by datasheets and to estimate the switching losses.

The measurements are conducted for all permutations of different parameters, i.e. different input voltages, various inductors and the two converter ICs fabricated in silicon and GaN, respectively. The loss model calculates all losses for each component of the converters based on the datasheet values and the measurement results. The output of the loss model includes key performance values such as the power efficiency η over V_{in}. Additionally, an analysis of the power loss is provided to give insights into the loss contribution based on different mechanisms (such as conduction loss and switching loss) as well as the individual contribution of the components forming the converter. Thereby, the exact loss mechanisms are explained, allowing the results and insights gained with the buck converter of this work to be transferred to other GaN-based power converter systems.

M. P. Kaufmann and B. Wicht, *Monolithic Integration in E-Mode GaN Technology*,
Synthesis Lectures on Engineering, Science, and Technology,
https://doi.org/10.1007/978-3-031-15625-0_5

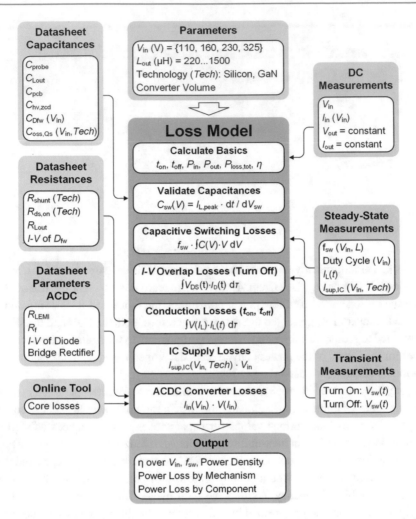

Fig. 5.1 Visualization of the developed loss model based on datasheet parameters and different measurements

The switching losses caused by recharging of (parasitic) capacitances as well as I-V overlap losses during switching transitions are examined in Sect. 5.2. The conduction losses are discussed in Sect. 5.3 before a detailed loss analysis is provided. Section 5.4 investigates, how the GaN technology can be exploited in order to increase power density while maintaining high efficiency. Thereby, a base for optimizing GaN power converters is provided in order to take full advantage of the technology benefits.

5.1   Monolithic Silicon Offline Buck Converter for LED Lighting

Power supply converters for LED lighting have become a large market, especially since the traditional incandescent light bulbs have been banned in many countries all around the world. Unlike the traditional light bulb, LEDs cannot be operated directly from the AC mains but require a power converter to provide a constant DC current. This converter requires high power density to fit inside a light bulb casing together with the LEDs. Additionally, high conversion efficiency is necessary due to the temperature sensitivity of LEDs and the limited cooling capabilities of standard lamps with light bulb sockets. For the mass market of home lighting, a high level of integration is desired in order to minimize the number of external components and associated cost for both the components and the PCB assembly.

This section presents and compares commercially available silicon converter ICs for LED lighting. After a chip level comparison between one silicon IC and the GaN IC of this work, the silicon IC is characterized on a high level and initial results are compared with respect to the performance of the GaN converter presented in Sect. 4.5.

Offline Buck Converter ICs

For the application space of LED lighting, various converter ICs with integrated power transistor, gate driver, control loop, supply generation, and protection circuits are commercially available, such as [1] and [2]. In the following part, they are compared on IC level to the GaN design of this work. Table 5.1 provides a summary of the feature set of two silicon converter ICs and also of the converter IC presented in Chap. 4.

All three converters are typically configured as low-side buck converter, operate in QRM, and employ an external shunt resistor for current sensing. They are suitable for operation directly from the rectified mains voltage and support an output power of at least 15 W, which is the typical power level of an LED lamp replacing a 100 W incandescent light bulb. Reference [2] uses a transformer to provide an isolated output. An additional auxiliary winding is required for both, the generation of the IC's supply voltage and the zero-current detection as part of the control loop to trigger a turn-on. Reference [1] and this work use a different approach and employ a basic inductor for a non-isolated output. Since an LED light bulb replacement is a fully cased system, no isolated output is required. For zero-current detection, both converters use a capacitive edge detection scheme at the switching node to trigger a turn-on (see Sect. 4.3). Additionally, they employ an integrated linear regulator to generate the supply voltage for the IC from the high-voltage switching node connected to the drain of the power transistor. Due to the similar operation conditions between this work and [1], these two converters are used for a detailed loss analysis and comparison in the following sections.

Figure 5.2 shows the die micrographs of the GaN buck converter IC and the silicon IC [1]. Due to the larger breakdown electrical field of GaN, the high-voltage power transistor

Table 5.1 Feature comparison of monolithic converter ICs in silicon and GaN technology

Reference	[1]	[2]	This work [3]
Year	2016	2014	2020
Technology	Si	Si	GaN
Operation mode	QRM	QRM	QRM
Current sense	Shunt	Shunt	Shunt
Zero-current detection	External capacitor	Auxiliary winding	Integrated capacitor
Integrated startup/supply	Yes	Yes	Yes
Protection*	UVLO, OTP, OCP	UVLO, OCP	UVLO, OCP
Isolated output	No	Yes	No
Max. P_{out}	22 W	15 W	29 W
BVDSS	725 V	800 V	650 V
$R_{DS,on}$	3.4 Ω	6.0 Ω	1.0 Ω
Die area	5.8 mm^2	5.5 mm^2	2.1 mm^2

*UVLO: undervoltage lockout, OTP: over temperature protection, OCP: over current protection

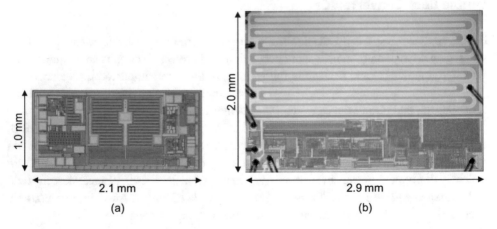

Fig. 5.2 Die micrographs: **a** GaN buck converter IC of this work, **b** silicon-based buck converter IC [1] drawn to scale

depicted in the center of Fig. 5.2a requires a smaller area and shows smaller feature sizes of the drain and source fingers than the power transistor in silicon (upper part in Fig. 5.2b). This becomes even more significant when the different $R_{DS,on}$ is considered, which is more than three times higher for the silicon IC compared to the GaN implementation (see Table 5.1). Additionally, the GaN IC integrates a high-voltage linear capacitor for zero-current detection to operate in QRM while the silicon-based converter requires a discrete high-voltage capacitor and resistor on PCB for this functionality. In summary, the GaN IC demonstrates much higher power density and integrates more high-voltage functions on a smaller die area. However, for power converter applications, the system power density and efficiency investigated further below are more important.

Top-Level Characterization of the Silicon Buck Converter

In order to compare the silicon and GaN-based converter ICs on the system level, the silicon IC is implemented as a low-side buck converter for the same application space as the GaN IC of this work (see Sect. 4.5). Similar external components are utilized to obtain comparable converter systems for both the silicon and the GaN IC. Figure 5.3 shows the populated PCBs of the LED power supply converters. A larger debug board depicted in Fig. 5.3a is used for a detailed characterization of the GaN-based converter. It provides access to internal signals such as the gate of the power transistor and supports different test modes to characterize individual subcircuits.

A second PCB is designed as an application board, which is optimized for the highest power density. The bottom view of this board in Fig. 5.3b includes the input rectifier, low equivalent series resistance (ESR) and low equivalent series inductance (ESL) ceramic input capacitors as well as the freewheeling diode, the shunt resistor, and the buck converter IC (see also the block diagram in Fig. 4.7). The compact PCB has a width of 22.9 mm and a

Fig. 5.3 Photographs of the populated buck converter PCBs: **a** Debug board for various test modes providing access to internal signals, **b** bottom view of an application board optimized for highest power density, **c** three-dimensional view of the application board

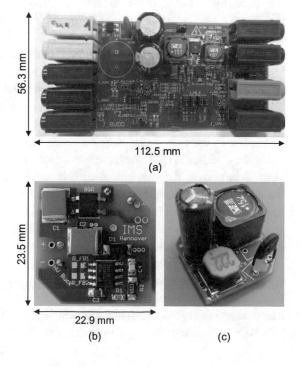

(a)

(b)

(c)

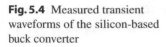

Fig. 5.4 Measured transient waveforms of the silicon-based buck converter

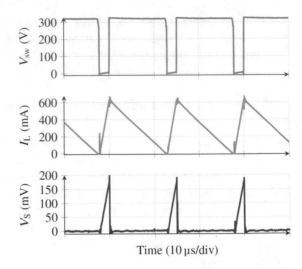

Time (10 μs/div)

height of 23.5 mm. Figure 5.3c shows a three-dimensional view of the top side of the PCB. It shows that the total volume of the converter is dominated by the output inductor and the bulky capacitor required to buffer the zero crossing of the line voltage. The capacitor size is defined by the converter power level and the input specifications regarding peak voltage and line frequency according to Eq. 4.11. Its size depends on the application and cannot be influenced by the design of the converter IC. This is different from the other large component, the output inductor. Its size can be reduced by increasing the switching frequency of the converter, leading to higher power density.

Figure 5.4 shows the top-level waveforms of the buck converter in Fig. 5.3. It is configured with $L_{out} = 1000\mu H$ leading to a switching frequency of 67 kHz. For this measurement, the input voltage V_{in} is at 325 V, which is the nominal peak voltage of the 230 VAC power grid. Due to similar operating conditions, the waveforms here resemble the ones for the GaN buck converter in Fig. 4.24.

Similar to the GaN-based buck converter characterized in Sect. 4.5, the silicon-based converter operates in QRM with cycle-by-cycle peak current control. When the power transistor is turned on, the switching node is at low level and the inductor current I_L rises. At the desired peak current of 635 mA, the current sense voltage V_S is at 184 mV triggering the power transistor to turn off. After that, the converter enters the freewheeling phase and I_L decreases. It goes to a minimum value of -5 mA causing a falling slope at V_{sw} which is detected by the controller IC and the power transistor is turned on again.

Figure 5.5 depicts the measured efficiency for both, the silicon and the GaN-based buck converters over the nominal V_{in} range covering the 110 V and 230 V AC line voltages. Figure 5.5a shows the results for a large inductor with 1000 μH inductance in a 12 × 12 × 10 mm package while Fig. 5.5b shows the efficiency values for a smaller 470 μH inductance in a 10 × 10 × 6 mm package. In both cases, the GaN-based converter achieves superior

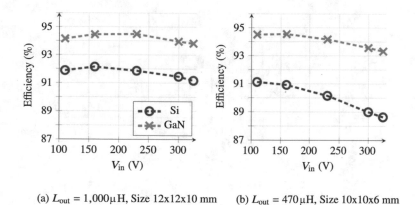

(a) L_{out} = 1,000μH, Size 12x12x10 mm   (b) L_{out} = 470μH, Size 10x10x6 mm

Fig. 5.5 Measured efficiency for GaN and silicon-based buck converters over V_{in} for two different coils

efficiency, which is even more significant for the smaller inductor size. There, the silicon-based converter shows an efficiency below 90% for input voltages above 250 V.

The GaN converter IC of this work achieves a much higher power density and integrated more high-voltage functions on a three times smaller die area than a commercially available silicon IC. Initial characterizations indicate superior efficiency of the GaN-based converter. In order to understand which characteristics of the GaN IC and mechanisms of the system implementation lead to the efficiency benefit, a detailed loss analysis is provided in the following sections.

5.2 Modeling of Switching Losses

In this section, the switching losses of both the GaN and the silicon-based converters are investigated. This is required to obtain insights into the loss mechanisms and contributions of the examined power converters. Therefore, the capacitance at the switching node is investigated based on transient measurements as well as datasheet parameters and manufacturer simulation models of the different components (see Fig. 5.1). Additionally, the I-V overlap losses during switching transitions are estimated from the characterization results by comparison between modeled and measured transient curves. The results are further used in Sect. 5.3 for a detailed loss comparison and for identifying dominant loss mechanisms and contributors.

In the following part, the capacitances at the switching node are examined. With power transistor, output inductor, freewheeling diode, and zero-current sense capacitor, there are multiple components connected to this node. Each of them adds some capacitance to the switching node, which has to be recharged at each switching transition when the power transistor turns on or off. In general, capacitive switching losses are proportional to the

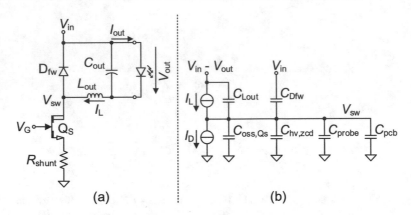

Fig. 5.6 **a** Simplified buck converter schematic; **b** equivalent capacitance net during slewing of V_{sw}

capacitance value C, the switching frequency f_{sw} and the difference of the squared voltages $V_{t2}^2 - V_{t1}^2$ before and after the charging. Thus, they are expected to contribute significantly to the power loss of offline converters with input voltages typically up to 325 V.

Figure 5.3a shows a simplified schematic of a low-side buck converter (see Fig. 4.7). When the transistor turns off, the inductor L_{out} forces the current I_L to flow into the switching node V_{sw} charging the capacitances at that node until the voltage V_{sw} is greater than V_{in} and the freewheeling diode D_{fw} becomes conducting (Fig. 5.6).

From the schematic in Fig. 5.3a, the equivalent capacitance net in Fig. 5.3b is derived. Due to the relatively low values of the transistor $R_{DS,on}$ as well as the current sense resistance R_{shunt}, they are negligible for the capacitance model. The current source I_D represents a drain current of Q_s, which may flow during slewing of V_{sw}. In the first step, the drain current is assumed to be zero during the switching transition of V_{sw}. I_D and its influence on the switching losses are investigated further below.

For the equivalent capacitor net, the parasitic and intended capacitances of all components connected to the switching node are considered. L_{out} itself has a parasitic inter-winding capacitance C_{Lout} between V_{sw} and V_{out}. The freewheeling diode D_{fw} connected between V_{sw} and V_{in} adds the capacitance C_{Dfw} of its intrinsic depletion region. Most capacitances, however, are located between V_{sw} and ground. They consist of the transistor output capacitance $C_{oss,Qs}$, the zero-current detection capacitor $C_{hv,zcd}$ and parasitic capacitances of the assembly summarized as C_{pcb}. This includes the capacitance of the bond pad on die as well as the pad and trace capacitances on the PCB. When the switching node is measured, the utilized oscilloscope probe also adds some capacitance C_{probe} to the switching node.

Switching Node Capacitance of the Silicon Buck Converter

In the following part, the capacitance C_{sw} at the switching node is extracted from the turn-off transient of the switching node voltage V_{sw}. Right after the power transistor Q_s is turned off, the inductor charges C_{sw} with an approximately constant current $I_{L,peak}$, assuming that I_D is negligible. Hence, the voltage-dependent switching node capacitance can be extracted from the transient slope of V_{sw} according to Eq. 5.1.

$$C_{sw} = I_{L,peak} \cdot \frac{1}{dV_{sw}/dt} \tag{5.1}$$

Figure 5.7a shows the transient at V_{sw} when the silicon power transistor turns off. In total, it takes about 45 ns from turn-off of the transistor until the switching node reaches 325 V. The varying slope represents the voltage dependence of the total capacitance at the switching node C_{sw}. For voltages above 150 V, V_{sw} shows the steepest slope of 14.4 V/ns with a linear characteristic. At lower voltages, the slope is smaller, which corresponds to a higher, voltage-dependent capacitance. This indicates that the non-linear $C_{oss,Qs}$ of the high-voltage silicon power transistor is dominant especially at low V_{sw}. When V_{sw} approaches the input voltage V_{in}, the voltage-dependent junction capacitance C_{Dfw} of the freewheeling diode increases. Thus, the slope of V_{sw} decreases again at the end of the switching transition.

Figure 5.7b shows the V_{sw} turn-off transient for the GaN-based buck converter with a much more uniform slope indicating a more linear total switching node capacitance C_{sw}. The voltage-dependent output capacitance $C_{oss,Qs}$ of the GaN transistor is not dominant. The total transition time from turn-off of the power transistor until the switching node reaches 325 V is 35 ns and thereby considerably lower than for the silicon-based converter. Similar published waveforms for the turn-off transient of GaN and silicon transistors [4] confirm the validity of this measurement. In order to characterize slew rates in the range of some ten

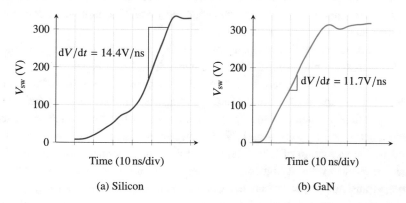

Fig. 5.7 Measured V_{sw} transient at turn-off for **a** silicon-based and **b** GaN-based buck converter

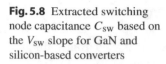

Fig. 5.8 Extracted switching node capacitance C_{sw} based on the V_{sw} slope for GaN and silicon-based converters

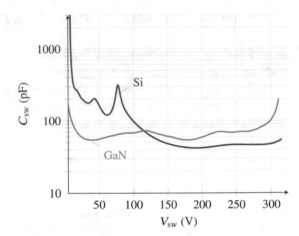

V/ns, a 500 MHz probe is used. The use of such a probe leads to a delay of approximately 300 ps while the slope of the switching node is precisely measured.

At turn-off of Q_s, the inductor current is at 635 mA (see Fig. 5.4). Sourced by an inductance of 1,000 μH, this current can be considered constant during slewing of V_{sw}. In the first step, it is assumed that Q_s is fully turned off during slewing of V_{sw} such that I_L is the only current present. Hence, the total value of the voltage-dependent capacitance C_{sw} at the switching node can be calculated based on the transient slope of V_{sw} according to Eq. 5.1. Figure 5.8 shows this calculated C_{sw} over V_{sw} for both the silicon and the GaN-based converters. Especially at low voltages, the capacitance in the silicon converter, is much larger than in the GaN-based converter. At higher voltages C_{sw} is slightly lower in the silicon implementation due to a different inductor and freewheeling diode used for the slower switching silicon converter.

According to Fig. 5.6b, the total capacitances C_{sw} are formed by parasitic capacitances of different components connected to the switching node. The individual contribution of each component is analyzed in the following.

Most capacitances are either independent of the DC bias or show a decrease of the capacitance with increasing bias voltage. Voltage independent, linear capacitances are typically capacitances formed by metal electrodes and air or other voltage-independent dielectrics. Examples of such capacitances are the inter-winding capacitance C_{Lout} or the capacitance C_{probe} of specifically tuned oscilloscope probes. In contrast to that, junction capacitances in semiconductors are non-linear and show a lower capacitance at a higher DC bias. This is caused by the formation of the depletion region, which increases the distance between the capacitance electrodes at higher voltage, leading to a lower capacitance. This applies for both the transistor output capacitance $C_{oss,Qs}$ and the diode capacitance C_{Dfw}. $C_{oss,Qs}$ is located between V_{sw} and ground and decreases with increasing V_{sw} explaining the negative slope of C_{sw} in Fig. 5.8 for $V_{sw} < 30$ V. In contrast to that, C_{Dfw} is connected between V_{sw} and V_{in}. Thus, it experiences a lower DC bias voltage for higher V_{sw}, what explains the positive slope of C_{sw} for $V_{sw} > 300$ V. Between 25 V and 180 V, the curve of the extracted C_{sw} shows parts where its value increases with higher voltage V_{sw}. This behavior cannot be explained by the capacitances and components connected to the switching node according to Fig. 5.6 (b). In contrary, it is an indication that Q_s is partially turned on during slewing of V_{sw} due to Miller coupling and a non-negligible drain current I_D is present. This is investigated further below.

In the first step, the distribution of C_{sw} across the different components is investigated for the silicon-based converter. The capacitances of the components can be extracted from the respective datasheets. The voltage-dependent distribution is illustrated in Fig. 5.9a. C_{Lout} is a voltage-independent capacitance of 15.1 pF [5] and the oscilloscope probe adds 9.5 pF [6]. Both $C_{oss,Qs}$ and C_{Dfw} show a voltage dependency as depicted in Fig. 5.9a. $C_{hv,zcd}$ is a ceramic capacitor of 100 pF. According to the application notes in [1], $C_{hv,zcd}$ is partially decoupled with a series resistance of 420 kΩ between its bottom plate and ground. Hence the effective capacitance during the fast transient of V_{sw} is reduced to 100 fF. It can thereby be neglected for the capacitance extraction from the slope of V_{sw} but it has to be considered for the overall switching loss of the converter. C_{pcb} is assumed to be 2.5 pF in order to achieve good matching between modeled and extracted total capacitance in the voltage-independent region of C_{sw} at $V_{sw} \sim 180$ V. This value is verified by coarse estimations of the PCB traces and land patterns. The pad of the output inductor with the routing between inductor and converter IC covers an area $A \sim 35$ mm. Using the basic relation for the capacitance of a plate capacitor (Eq. 5.2) with $\epsilon_r = 4.5$ and $d = 1.55$ mm for the utilized FR4 PCB, this leads to a capacitance of 800 fF of this part alone.

$$C_{plate} = \epsilon_0 \cdot \epsilon_r \cdot \frac{A}{d} \tag{5.2}$$

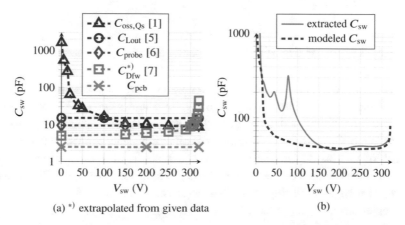

(a) *) extrapolated from given data

(b)

Fig. 5.9 Silicon-based buck converter: **a** Distribution of the voltage-dependent C_{sw} over different components according to datasheets [5–7], **b** sum of modeled capacitances and characterized total C_{sw}

A similar PCB area is utilized to connect the freewheeling diode to the switching node. Additionally, the pins of the diode and the lead of the converter IC as well as the solder connection contribute to the value of C_{pcb}. Thus, a value greater than 1.6 pF is expected, validating the assumed 2.5 pF.

Figure 5.9b shows the measured value of C_{sw} over V_{sw} by a solid line. It is extracted from the slope of V_{sw} according to Eq. 5.1. The dashed line shows the value of the modeled capacitance over V_{sw} as the sum of all capacitances depicted in Fig. 5.9a. For voltages below 15 V and above 150 V, the modeled capacitance shows good accordance with the measured C_{sw} extracted from the slope of V_{sw}. However, between 15 V and 150 V, the extracted C_{sw} is much higher than the sum of the capacitances in Fig. 5.9a represented by the dashed line in Fig. 5.9b. The difference between the two curves as well as the intermittent rising slopes indicate, that C_{sw} is not charged with a constant current, but a part of I_L flows through the transistor Q_s. This is likely caused by a partially turning on of Q_s caused by capacitive Miller coupling from the switching node into the gate node. The drain current I_D of Q_s during switching transitions is investigated further below.

In the previous part, the capacitance C_{sw} at the switching node is estimated based on the voltage slope of the turn-off transition as well as values provided in datasheets and simulation models (see Fig. 5.1). At the turn-off transition, C_{sw} is charged with a current source characteristic by the output inductor L_{out}. Thus in a first-order approximation, this charging can be assumed to be lossless. Nevertheless, the charging of C_{sw} by L_{out} slightly increases the RMS value of the inductor current and the associated conduction losses. This is captured by a measurement of the inductor current and modeled as part of the conduction losses in Sect. 5.3. In contrast to that, the discharging of C_{sw} happens across the channel resistance of the power transistor Q_s during the turn-on transition leading to capacitive discharging losses. This is examined further below based on the voltage-dependent value of C_{sw} characterized above.

I-V Overlap Losses of the Silicon Buck Converter

In this part, the I-V overlap losses of the silicon buck converter are estimated for the turn-off transition of the power transistor. The plot in Fig. 5.9b shows two curves for the switching node capacitance C_{sw} of the silicon-based buck converter. The dashed line represents the sum of the capacitances taken from the datasheets of all components connected to the switching node. Each of them is depicted in Fig. 5.9a. C_{sw} consists of a voltage independent part related to the inter-winding capacitance of the output inductor as well as PCB traces and the oscilloscope probe. Additionally, it shows a decreasing slope for low voltages associated with the depletion output capacitance of the power transistor, as well as an increasing slope for high voltages associated with the junction capacitance of the freewheeling diode. The solid line in Fig. 5.9b represents a value for C_{sw} extracted from the slope of V_{sw} according to Eq. 5.1 assuming a constant current $I_{L,peak} = 635$ mA charging the capacitances. However,

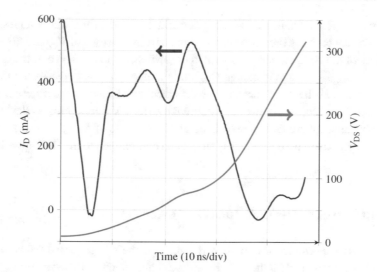

Fig. 5.10 Overlap of estimated I_D and measured V_{DS} during turning of the silicon-based Q_s

the progression of this curve with intermittent rising slopes cannot be explained by any of the capacitances identified to be connected at this node. Thus, the assumption that the full current $I_{L,peak}$ charges C_{sw} has to be reconsidered. More likely, the capacitive Miller coupling from the drain into the gate of the power transistor during slewing of V_{sw} is sufficient to partially turn-on Q_s. Hence, a part of the inductor current is dissipated as I_D through the transistor and not used to charge C_{sw}. Assuming that the modeled values for C_{sw} in Fig. 5.9b are correct, I_D can be calculated using the relations given in Eq. 5.3 based on the difference between the modeled and the extracted values of C_{sw}.

$$I_D = I_{L,peak} - \left(C_{sw,extracted} - C_{sw,modeled}\right) \cdot \frac{dV_{sw}}{dt} \tag{5.3}$$

The thereby estimated transient drain current I_D of Q_s is depicted in Fig. 5.10 together with $V_{DS} \sim V_{sw}$. During slewing, a considerably high drain current of up to 525 mA flows through Q_s while V_{DS} is already above 70 V. This leads to I-V overlap losses as high as 40 W for some nanoseconds. The overlap energy can be calculated as the integral over time of the I-V product during slewing. For the silicon-based converter, the overlap energy is approximately 635 nJ at each turn-off of Q_s.

In order to develop a loss model for the silicon-based buck converter, the voltage-dependent capacitance of the switching node C_{sw} has been estimated. In a first step, it is extracted from the slope of the switching node voltage V_{sw} with the assumption, that C_{sw} is charged with a constant current $I_{L,peak} = 635\text{mA}$. To verify this approach, the thereby extracted C_{sw} is compared with the capacitances added to C_{sw} by all components connected to this node. Both methods show good matching of the estimated C_{sw} for voltages above

170 V. Thus, the general approach seems to be valid. However, for lower voltages, there is a large discrepancy between the C_{sw} values extracted from the slope of V_{sw} and the sum of the capacitances of the components connected to the switching node. Hence, it is assumed that only a part of $I_{L,peak}$ is used to charge C_{sw} while most of $I_{L,peak}$ is dissipated through the transistor channel as the drain current I_D. I_D is calculated based on the difference between modeled and extracted values of C_{sw}. Additionally, the I-V overlap losses during turn-off of the power transistor are estimated by integration over time of $I_D(t) \cdot V_{DS}(t)$. The results of this investigation are used for the loss model and comparison to the GaN-based converter in Sect. 5.3.

Switching Losses of the GaN Buck Converter

According to the investigations above performed for the silicon-based converter, similar investigations are executed for the GaN-based buck converter. In a first step, the capacitances forming C_{sw} depicted in Fig. 5.6b are estimated. After that, the drain current I_D during slewing of V_{sw} is calculated. Based on the capacitance modeling for the silicon converter in Fig. 5.9, it is assumed that the datasheet values for C_{Dfw}, C_{pcb}, C_{probe} and C_{Lout} are correct. In contrary to the silicon converter, the GaN IC integrates the zero-current detection capacitor $C_{hv,zcd} = \sim 100$ fF on die. It is intrinsically coupled to the drain of Q_s and is therefore considered to be a part of $C_{oss,Qs}$.

Figure 5.11a shows the voltage-dependent distribution of C_{sw} on the different components according to Fig. 5.6b for the GaN-based converter. For the IC designed in this work, no datasheet exists which would provide values for the power transistor output capacitance $C_{oss,Qs}$. Thus, the voltage depended value for this capacitance has to be estimated based on

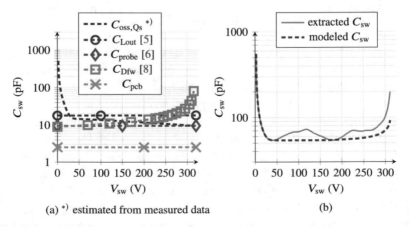

(a) *) estimated from measured data   (b)

Fig. 5.11 GaN-based buck converter: **a** Distribution of the voltage-dependent C_{sw} on different components according to datasheets [5, 6, 8], **b** sum of modeled capacitances and characterized total C_{sw}

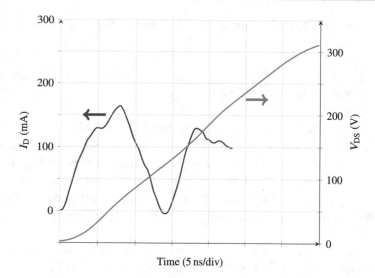

Fig. 5.12 Overlap of estimated I_D and measured V_{DS} during turning of the GaN-based Q_s

measurements. As discussed above, this capacitance is a depletion capacitance and therefore shows a reduced value at higher bias voltage V_{sw}. Hence, it is responsible for the decreasing slope of the curve for C_{sw} in Fig. 5.8b (see also Fig. 5.11b "extracted C_{sw}"), which is extracted from the voltage slope of V_{sw} during turn-off according to Eq. 5.1. The value for $C_{oss,Qs}$ at high DC bias is estimated by comparing the sum of C_{Dfw}, C_{pcb}, C_{probe} and C_{Lout} from Fig. 5.11a to the total value of C_{sw} extracted from the slope of V_{sw} in Fig. 5.8b. This comparison is performed at the local minimum of C_{sw} at $V_{sw} = 180$ V where both depletion capacitances of transistor and diode show only a weak voltage dependence.

Figure 5.11b depicts the modeled C_{sw} together with the extracted values from the V_{sw} slope in Fig. 5.7b. Similar to the silicon-based converter, the voltage dependency of C_{sw} is caused by the two depletion capacitances $C_{oss,Qs}$ and C_{Dfw}. As expected from the technology basics, the GaN transistor in Fig. 5.11a shows a much lower $C_{oss,Qs}$ compared to the capacitance of the silicon transistor depicted in Fig. 5.9a. The GaN transistor shows values of 40 pF at 10 V and 15 pF at 50 V, while the silicon transistor shows 250 pF and 30 pF, respectively.

Similar to the investigation performed for the silicon-based converter, it is assumed that the difference between modeled and extracted is caused by a drain current I_D. Figure 5.12 shows the estimated drain current based on Eq. 5.3 together with the measured V_{DS} during slewing of V_{sw} for the GaN power transistor.

With less than 180 mA, the estimated I_D for GaN is considerably smaller than for the silicon-based transistor (see Fig. 5.10). This leads to smaller I-V overlap losses, which can be integrated as $I_D(t) \cdot V_{DS}(t)$ over the transition time to approximately 335 nJ. This energy is dissipated in the transistor channel at each turn-off of Q_s. This value is further utilized in Sect. 5.3 for a detailed loss analysis of the full buck converter and a comparison to the silicon converter.

Losses at the Turn-On Transition

This part investigates the switching losses during the turn-on transition of the power transistor Q_s. Since the converters operate in QRM, the turn-on occurs at approximately zero inductor current. Thus, the I-V overlap losses are negligible during turn-on of Q_s. However, the switching node capacitance C_{sw} is discharged across the transistor channel resistance and contributes to the switching losses. This is in contrast to the turn-off transition of the previous investigation, where C_{sw} is charged lossless by the output inductor with a current source characteristic.

Figure 5.13 shows the measured V_{sw} transient at turn-on of Q_s for the silicon converter. This corresponds to Figure 4.19 of the GaN-based converter. The slightly negative slope shows the time when the inductor current I_L reverses and starts to discharge C_{sw} with a current source characteristic. Additionally, the reverse recovery charge of the freewheeling diode is dissipated in a lossless, resonant manner.

After approximately 300 ns (marked as t_1 in Fig. 5.13), the silicon IC detects the falling slope and turns on Q_s. This is within the same range as for the GaN-based converter, which triggers the turn-on after 280 ns. After that, C_{sw} is rapidly discharged with a slope as steep as -13.3 V/ns. This slew rate is approximately 40% higher than for the GaN-based transistor which is caused by the stronger pull-up of a gate driver integrated in CMOS technology (see Sect. 3.3).

The capacitive discharging losses are dissipated in the transistor channel and are independent of the slew rate of V_{sw} or the resistance value of the transistor channel. However, thanks to the operation in QRM, not the complete energy stored on C_{sw} has to be dissipated

Fig. 5.13 Transient V_{sw} during turn-on of the silicon-based converter

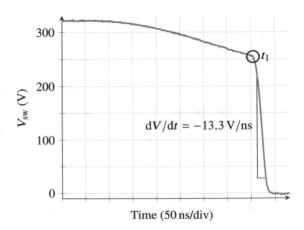

Time (50 ns/div)

across a resistance. The energy dissipated through the transistor channel can be calculated according to Eq. 5.4 utilizing $C_{sw}(V)$ as given in the plots in Figs. 5.9 and 5.11, respectively. The limits of the integral are $V_1 = 0$ V and V_2 is the voltage when the steep slope in Fig. 5.13 begins. For the silicon converter, it is approximately 70 V below V_{in} and for the GaN converter, it is around 60 V below V_{in} (see Fig. 4.19).

$$E_{Csw} = \int_{V_1}^{V_2} C_{sw}(V) \cdot V \, dV \qquad (5.4)$$

For the case depicted in Fig. 5.13 with $V_{in} = 325$V, the energy dissipated at each turn-on is $\sim 2.1\,\mu$J for the silicon converter and 1.2 mJ for the GaN converter, respectively.

In this Section, the switching losses for both the silicon and the GaN-based converters have been examined. Based on datasheet parameters, the voltage-dependent capacitance C_{sw} at the switching node is estimated and its distribution across the various components is modeled. Additionally, C_{sw} is extracted from the slope of the switching node voltage V_{sw} at the turn-off transition. Both values for C_{sw} show good matching for higher voltage. The thereby validated capacitance value is used by the loss model (see Fig. 5.1) to calculate the capacitive switching losses of the converter at each turn-on of the power transistor Q_s. At turn-off, however, C_{sw} is charged lossless with a current source characteristic of the output inductor.

The modeled and extracted values for C_{sw} over V_{sw} show a large discrepancy at intermediate voltages between 25 V and 200 V, likely caused by a drain current through the power transistor during slewing of V_{sw}. Thus, the difference between the two values of C_{sw} obtained by different methods is utilized to estimate the drain current and additionally the I-V overlap losses at each turn-off of Q_s. Due to operation in QRM, the transistor is turned on at zero current. Thus, no I-V overlap losses occur at the turn-on of Q_s. The results of this section are the base for a detailed loss analysis and comparison between the silicon and the GaN implementation in Sect. 5.3, where also other loss mechanisms such as conduction losses are investigated and compared.

5.3 Loss Distribution

There are various sources for the power loss in a switched mode power converter. Therefore, a detailed loss analysis is valuable in order to understand loss mechanisms and optimize the converter towards high efficiency. Additionally, when the loss mechanisms are understood, the results of this work can also be transferred to other power converter systems.

This work considers offline converters operated directly from the AC mains. The first loss contribution occurs in the ACDC converter stage, which is composed of a diode bridge rectifier, an EMI filter, and an inrush current limiting NTC resistor. These losses depend mostly on the converter's input current and are independent of the subsequent buck converter.

The power dissipation of the converter IC itself represents another source of losses. The IC requires a supply voltage to bias all circuits integrated on die. Both the GaN and the silicon IC generate their own supply voltage directly from the high-voltage input utilizing a linear regulator. The supply current consumption of the GaN IC is investigated in Sect. 4.4 based on the ripple of the supply voltage buffered by an auxiliary capacitor on PCB. It is determined to be between 400 mA and 831 mA depending on the duty cycle of the converter. The same method is employed to determine the supply current consumption of the silicon converter, which is 1200 mA.

The remaining losses can be classified into two categories: switching losses, which scale with the operation frequency and conduction losses, which scale with the current. The switching losses are modeled according to the investigations in Sect. 5.2 based on Fig. 5.6. Additionally, the inductor core losses are estimated using an online tool provided by the coil manufacturer.

For the conduction losses, a DC loss model is developed. Since the current path leading to conduction losses is different for the on-phase of Q_s and the freewheeling phase, the model has to distinguish both phases. Additionally, the duration of the phases depends on V_{in} if the converter operates in QRM (see Sect. 4.1).

Conduction Loss Model

The following part investigates the contribution of different components in the power path to the conduction losses. The current path has to be distinguished for the two operation phases: the on-phase, where the high-voltage transistor Q_s is turned on to energize the inductor L_{out}, and the freewheeling phase where Q_s is turned off and the current flows through the freewheeling diode D_{fw}. The resistance values and the associated losses are then compared between the silicon and the GaN implementation.

From the simplified schematic of the buck converter in Fig. 5.14a, the equivalent circuit in Fig. 5.14b can be derived for the on-phase of Q_s. The rising slope of the triangular current I_L (see also Fig. 5.4) passes through the inductor with the DC copper resistance R_{Lout},

Fig. 5.14 **a** simplified schematic of the buck converter, **b** equivalent resistance net during the on-phase of Q_s, **c** equivalent conduction loss model during the freewheeling phase

the transistor channel with its on-resistance $R_{DS,on}$ and the current sense resistor R_{shunt}. Figure 5.14c shows the equivalent DC model during the freewheeling phase where the decreasing slope of I_L passes again through the inductor R_{Lout} as well as the freewheeling diode. The latter is modeled as DC voltage source V_{Dfw} with a differential series resistance r_{Dfw}. A transient measurement of I_L (see Fig. 5.4) is used to calculate the conduction losses. Most of the measured current I_L is provided to the load as low-pass filtered I_{out}. However, the RMS value of I_L also captures some additional ripple current utilized for inductive charging of the switching node capacitance C_{sw} during turn-off of Q_s. The additional ripple due to resonant dissipation of the freewheeling diode's reverse recovery charge as well as the partially resonant discharge of C_{sw} right before Q_s turns on (see Sect. 5.2) is also included in the measured I_L.

The fraction of time when the inductor current I_L flows through both the transistor and the current sense resistor is defined by the duty cycle of the converter $D = V_{out}/V_{in}$. Thus, the contribution of $R_{DS,on}$ and R_{shunt} to the conduction losses is more significant at lower V_{in} while the diode conduction losses are more dominant at higher V_{in}. The inductor DC resistance R_{Lout} caused by the finite conductivity of the utilized metal wire is most critical for the conduction losses, since it is always part of the current path. Additionally, the inductance value defines the operation frequency of a buck converter operated in QRM and thereby scales the switching losses. Thus, the selection of the power inductor is carefully considered.

Choice of Suitable High-Voltage Inductors

The investigations above suggest that the output inductor has a strong influence on the losses of the implemented power converters. On one hand, it is always in the current path and contributes to the conduction losses (see Fig. 5.14), on the other hand, it determines the switching frequency of the converters operating in QRM and thereby scales the switching losses. This part identifies power inductors suitable for the given application with input voltages of up to 400 V and peak inductor currents as high as 700 mA. Additionally, the trade-offs for the inductor selection regarding switching and conduction losses are discussed.

Table 5.2 gives an overview of parameters and parasitics of different 400 V rated inductors in the same package. The parasitic capacitance is mostly related to the package size and is thus nearly independent of the inductance value. However, the other parameters correlate directly with the inductance value. A lower inductance value achieves a higher saturation current I_{sat} and lower copper resistance R_{Lout} but leads to a higher switching frequency for the same input voltage and current ripple (see Eq. 4.8). The silicon converter IC has a minimum specified switching frequency of 18 to 22 kHz [1]. Hence, the inductors of this series with an inductance value of 2,200 μH and higher are not considered. Furthermore, the converter IC supports a minimum on-time of 1.1 μs for the integrated power transistor. At $V_{in} =$

Table 5.2 Comparison of 400 V rated inductors in a 12 × 12 × 10 mm package [5]

Sold as	220 μH	470 μH	680 μH	1,000 μH	1,500 μH	2,200 μH
Typ. inductance	191 μH	423 μH	622 μH	855 μH	1344 μH	1904 μH
I_{sat}	2.0 A	1.4 A	1.2 A	0.9 A	0.8 A	0.65 A
R_{Lout}	0.3 Ω	0.68 Ω	0.96 Ω	1.2 Ω	1.9 Ω	3.1 Ω
C_{Lout}	17.4 pF	15.1 pF	17.9 pF	15.1 pF	17.1 pF	18.0 pF
f_{sw} at $V_{in} = 110$ V	191.8 kHz	90.2 kHz	62.0 kHz	45.8 kHz	29.9 kHz	21.1 kHz
f_{sw} at $V_{in} = 325$ V	275.4 kHz	132.1 kHz	91.2 kHz	67.5 kHz	43.9 kHz	31.0 kHz

325 V where the duty cycle is 15%, this corresponds to a maximum switching frequency of 140 kHz. Hence, the coils of the investigated inductor series with an inductance value of 220 μH and below are outside the specified operation condition of the silicon converter IC.

The investigations performed for the GaN converter in Chap. 4 suggest that the control accuracy of the output current degrades with higher input voltages (see Fig. 4.27). The control accuracy is defined by the delays in the turn-on path and the turn-off path of the controller ICs with respect to the converter cycle time. Hence, a lower inductance value with a smaller cycle time leads to increased challenges for the output current accuracy. Therefore, the average output current over input voltage is characterized by the silicon-based converter assembled with the different inductors below 2200 μH listed in Table 5.2. The results are depicted in Fig. 5.15.

For the largest inductance value, the output current drops by 4.4% at $V_{in} = 325$ V. On contrary to that, the output current increases significantly for coils with lower inductance

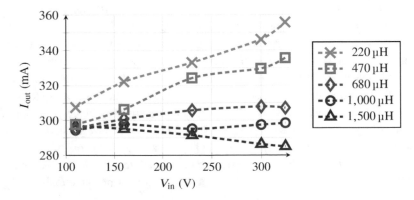

Fig. 5.15 DC output current over input voltage for the silicon converter with different coils

values. The combination of $L_{out} = 470\,\mu H$ and $V_{in} = 230\,V$ leads to a on-time of $1.9\,\mu s$ which is significantly larger than the specified max. value of the minimum on-time in the data sheet [1]. Despite that, the output current is 9% higher than at $V_{in} = 110\,V$ while the datasheet claims $\pm 3\%$ current control accuracy. The worst-case current deviation over V_{in} occurs at the combination of minimum inductance and maximum input voltage, where the output current is almost 16% higher than at $V_{in} = 110\,V$ for the same inductor. Since the $220\,\mu H$ inductor leads to an operation frequency outside of the specified capabilities of the silicon IC and to a significantly impaired current control, this inductor is not used for the following loss analysis of the silicon-based buck converter. The other inductors are employed to change the operation frequency of the converter in order to analyze the trade-off between switching losses and conduction losses further below.

Loss Distribution by Mechanisms

This part examines the losses in the converters by fundamental mechanisms: ACDC conversion losses, conduction losses inside the buck converter, switching losses in the buck converter, and supply losses for the converter IC. Figure 5.16a shows the loss distribution for the silicon-based buck converter when operated at a nominal line voltage of North- and Central America power grid $V_{in} = 110\,V$. Figure 5.16b shows the direct comparison when the buck converter is operated at $V_{in} = 325\,V$, which is the nominal peak voltage of the 230 V power grid. These two voltage values are chosen to cover a wide voltage range within the typical operation conditions for offline power converters. The different switching frequencies are achieved by employing the different inductors of Table 5.2 suitable for the silicon buck converter.

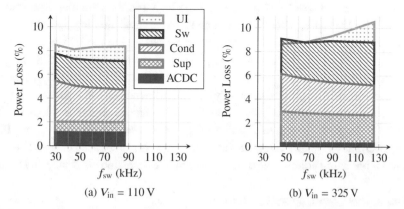

(a) $V_{in} = 110\,V$ (b) $V_{in} = 325\,V$

Fig. 5.16 Loss distribution for the silicon-based buck converter for **a** $V_{in} = 110\,V$ and **b** $V_{in} = 325\,V$

The losses of the ACDC input converter ("ACDC" in Fig. 5.16) are independent of the buck converter's operation frequency and therefore constant over f_{sw} in each of the two plots. However, at constant power the input current I_{in} leading to losses in the ACDC conversion stage is proportional to $1/V_{in}$. Thus, the ACDC losses reduce with higher V_{in} when a lower current I_{in} passes through the components. The ACDC losses are composed of the diode bridge rectifier losses $P_{loss} = V_f \cdot I_{in}$, the losses of the inrush current limiting resistor $P_{loss} = R_f \cdot I_{in}^2$ and the DC-resistance losses of the EMI filter inductor $P_{loss} = R_{L,EMI} \cdot I_{in}^2$. Hence the ACDC losses scale even stronger than with $1/V_{in}$.

The supply current of the silicon converter IC ("Sup" in Fig. 5.16) is nearly independent of the buck converter's operation point. Since the supply voltage for the IC is generated directly from the high-voltage switching node during the phase when $V_{sw} \sim V_{in}$, the supply losses are proportional to V_{in} and increase significantly at higher input voltage.

The conduction losses ("Cond" in Fig. 5.16) become smaller with higher switching frequencies. This is caused by the lower R_{Lout} of the inductors with lower inductance values, which leads to the higher switching frequency (see Table 5.2). The conduction losses also reduce with higher V_{in}. This is caused by the reduced duty cycle leading to a lower contribution of $R_{DS,on}$ and R_{shunt} to the conduction losses (see Fig. 5.14).

At $V_{in} = 110\,V$, the switching losses ("Sw" in Fig. 5.16) increase only slightly with higher switching frequency. Due to the QRM operation mode, most of the energy stored in the capacitance at the switching node is resonantly dissipated. Hence, the switching losses in general are not dominant at lower input voltages. At $V_{in} = 325\,V$ however, the switching losses are higher and increase even further with higher operation frequency.

The last remaining loss fraction is labeled "UI" in Fig. 5.16 and corresponds to unidentified losses. They are calculated as the difference between measured total power loss and the total loss predicted by the loss model. Negative values for the unidentified losses correspond to a loss overestimation by the model, which is represented by the gray outline cutting through the red area "Sw". This occurs in Fig. 5.16b for switching frequency values below 65 kHz. Otherwise, when the gray line is above the switching loss area, the model underestimates the total losses. The losses not captured by the model are likely related to self-heating effects of the inductor and the power transistor, especially at high input voltage and switching frequency. Nevertheless, if the achievable converter efficiency shall be predicted, the unidentified losses have only a small influence on the result.

This characterization of the silicon-based buck converter suggests a loss optimum if the $1,000\,\mu H$ inductor is utilized, which leads to $f_{sw} = 45.8\,kHz$ at $V_{in} = 110\,V$ and $f_{sw} = 67.5\,kHz$ at $V_{in} = 325\,V$, respectively. This result is further used in Sect. 5.4 to discuss the optimum operating points in order to achieve high conversion efficiency and high power density.

A similar loss investigation is performed for the GaN-based converter. The distribution of losses is depicted in Fig. 5.17a for $V_{in} = 110$ V and in Fig. 5.17b for $V_{in} = 325$ V. Since the GaN converter is able to support higher switching frequencies and shorter on-times, also the $220\,\mu H$ inductor can be included in the investigation, which is not possible for the silicon

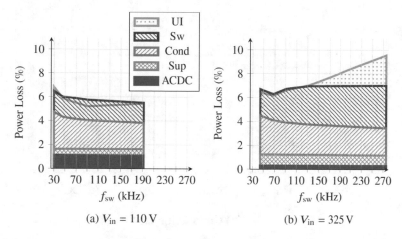

Fig. 5.17 Loss distribution for the GaN-based buck converter for **a** $V_{in} = 110\,V$ and **b** $V_{in} = 325\,V$

converter. Thus, the plots in Fig. 5.17 show a wider frequency range than the plots for the silicon converter in Fig. 5.16.

The "ACDC" losses are similar for both the GaN and the silicon-based converters since they are operated at the same input conditions. As characterized in Sect. 4.4, the GaN converter IC consumes a lower supply current, leading to lower supply losses "Sup" compared with the silicon converter. Since the supply is also generated from the high-voltage switching node, the supply losses scale proportionally with V_{in}, accordingly. The GaN converter shows lower conduction losses "Cond" which is mainly achieved by the much lower transistor on-resistance $R_{DS,on}$ of $1\,\Omega$ instead of $3.4\,\Omega$ for the silicon device (see Table 5.1). Thus, despite the current sense resistor R_{shunt} applied for the GaN converter has a larger resistance value, the total resistance $R_{DS,on} + R_{shunt}$ and the associated conduction losses are smaller than for the silicon converter. Thanks to the lower output capacitance and lower I-V overlap losses of the GaN IC, it shows lower switching losses than the silicon IC. However, similar for both technologies, the switching losses increase with V_{in} and also with the switching frequency f_{sw}. The unidentified losses "UI" represent the difference between the measured overall system loss of the converter and the loss predicted by the implemented model. In general, there is a good accordance between measurement and model with a deviation smaller than 0.5% for most operating points, where a negative value of "UI" corresponds to a loss overestimation of the model.

Loss Distribution by Component

In order to identify the critical components for optimizing the converter efficiency, a detailed loss comparison is performed for both the silicon and the GaN-based converters. Thus, the switching and conduction losses depicted in Figs. 5.16 and 5.17 are further divided into the components where they occur. Both implementations are assembled with similar compo-

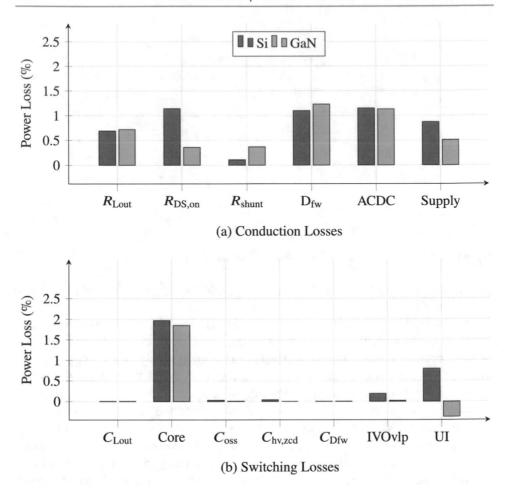

Fig. 5.18 Loss contribution of different components at $V_{in} = 110$ V and $L_{out} = 1,000\,\mu$H in a $12 \times 12 \times 10$ mm package: **a** conduction losses and **b** switching losses

nents. Due to the best matching between modeled and measured losses, the comparison is performed for the coil with $1,000\,\mu$H inductance. For the comparison, two different input voltages 110 V and 325 V are used.

Figure 5.18 shows the case for $V_{in} = 110$ V. The conduction losses depicted in Figs. 5.16 and 5.17 are further analyzed in order to obtain the distribution of the various components of the buck converters.

The results are compared between the silicon and the GaN-based implementation with the bar plot in Fig. 5.18a. The GaN buck converter achieves higher efficiency because a lower input power is required for the same output power. Thus, the losses of the ACDC bridge rectifier and EMI filter are slightly lower due to a lower input current. The most significant difference is the conduction loss due to the $R_{DS,on}$ of the transistor channel, which is much lower for the GaN-based converter. Additionally, the lower power consumption of the GaN IC leads to lower supply losses. Due to the larger R_{shunt} for the GaN implementation, the losses associated with the current sense are higher for this setup. Adding all conduction losses shown in Fig. 5.18a at $V_{in} = 110$ V for both implementations, the GaN converter shows a total of 4.3% conduction losses with respect to the input power compared to 5.1% for the silicon.

The switching losses of the converters are examined in Fig. 5.18b. Thanks to the QRM operation mode with partial resonant discharging, the capacitive losses related to C_{Lout}, C_{oss}, $C_{hv,zcd}$ and C_{Dfw} are very low and can be neglected. The dominant contribution to the switching losses is the core losses of the power inductor. They are estimated based on the measured operation frequency and inductor current ripple using an online tool provided by the coil manufacturer. A smaller but still significant contribution is caused by the I-V overlap losses during turn-off of the power transistor, where the silicon IC shows higher losses. Additionally, the loss model for the silicon converter underestimates the losses by 0.8% while the model for the GaN converter overestimates the losses by 0.4%. The sum of the conduction losses and the switching losses leads to the efficiency difference of 2.3% between the silicon and the GaN-based converters as depicted in Fig. 5.5a for $V_{in} = 110$ V.

Figure 5.19a shows the conduction loss distribution for $V_{in} = 325$ V. Due to the smaller duty cycle at higher V_{in}, the conduction losses of $R_{DS,on}$ and R_{shunt} are lower than for $V_{in} = 110$ V. However, ratio between the losses of the silicon and the GaN implementation stays the same. The lower input current at higher V_{in} with constant output power leads to reduced losses of the ACDC conversion. However, the relatively longer freewheeling phase leads to a significant increase in the diode conduction losses. Since the current always flows through the inductor, the conduction losses associated with the wire resistance R_{Lout} are nearly independent of V_{in}. In contrary to that, the supply losses scale linearly with V_{in}. Due to the larger supply current consumption of the silicon IC, this leads to a very significant loss contribution of 2.7%, which is three times higher than for the GaN IC.

The switching losses in Fig. 5.19b show again that the core losses of the power inductor are dominating. However, due to the capacitive losses being proportional to V_{in}^2, the losses related to C_{Lout}, C_{oss}, $C_{hv,zcd}$ and C_{Dfw} are more significant at $V_{in} = 325$ V and cannot be neglected anymore. Especially the lower transistor output capacitance C_{oss} of GaN as well as the smaller I-V overlap losses increase the efficiency benefit of the GaN-based converter. Deducing the overestimation of the loss model for the GaN implementation, this leads to a total loss of 6.2% while the silicon converter shows 8.8% at $V_{in} = 325$ V (see Fig. 5.5a).

5.4   Operation Point: Trade-Off Efficiency Versus Power Density

The investigations in Sect. 5.3 show that the choice of the inductor is critical for achieving high efficiency. This is due to the significant losses of the inductor core as well as the wire resistance which is always in the power path. Additionally, the inductance value affects the

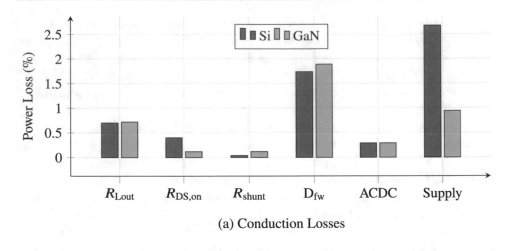

(a) Conduction Losses

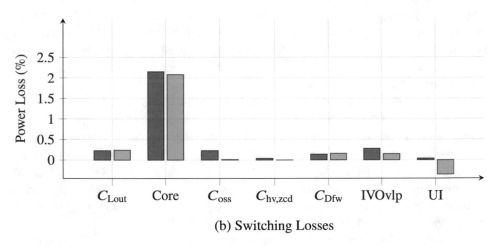

(b) Switching Losses

Fig. 5.19 Loss contribution of different components at $V_{in} = 325\,\text{V}$ and $L_{out} = 1,000\,\mu\text{H}$ in a $12 \times 12 \times 10\,\text{mm}$ package: **a** conduction losses and **b** switching losses

operation frequency of QRM buck converters and thereby scales the switching losses. For a constant package size, a lower inductance value leads to weaker magnetization of the inductor core which reduces the core losses. Furthermore, the shorter wire length required for a smaller inductance leads to a reduced DCR and smaller associated conduction losses. However, as the operation frequency increases, all capacitive switching losses as well as the core losses caused by changing magnetization increase. In order to identify the inductor achieving the highest efficiency, the silicon and the GaN-based converters are characterized by five different inductors between $220\,\mu\text{H}$ and $1,500\,\mu\text{H}$ (see Table 5.2). Since the input voltage of the converters affects the operation frequency, the duty cycle, and the capacitive switching losses, the characterization is performed for various input voltages. 110 V and

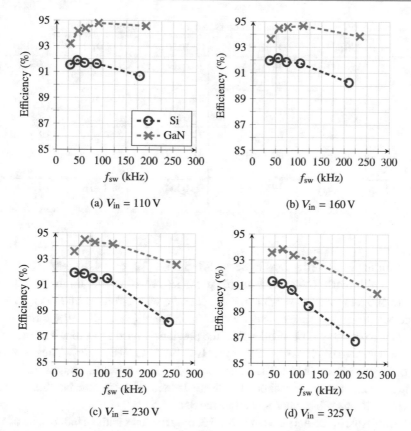

Fig. 5.20 Comparison of measured efficiency over switching frequency between GaN and silicon-based converters for **a** $V_{in} = 110$ V, **b** $V_{in} = 160$ V, **c** $V_{in} = 230$ V and **d** $V_{in} = 325$ V

160 V are chosen as the RMS value and nominal peak value of the power grid in North America. 230 V and 325 V are investigated as the RMS value and nominal peak value of the power grid in most other countries.

Figure 5.20 shows the measured efficiency of both converters over operation frequency. At all operating conditions, the GaN-based converter outperforms the efficiency of the silicon-based converter by at least 2% which corresponds to a loss reduction of 25%.

The GaN-based converter shows the best efficiency at higher operation frequencies and maintains a conversion efficiency greater than 93% over the full nominal input voltage range 110 to 325 V if the switching frequencies are limited to values below 130 kHz. This is achieved by the inductors with $L \geq 470\mu$H listed in Table 5.2. However, especially at high input voltages, the achieved efficiency reduces for higher operation frequencies above 130 kHz if the 220 μH inductor is utilized.

When higher power density is desired, the package size of the utilized inductor can be reduced. The 400 V rated inductors are available in three different package sizes. The largest

Table 5.3 Comparison of 400 V rated inductors in a $10 \times 10 \times 6$ mm package [5]

Sold as	220 μH	470 μH	680 μH	1,000 μH	1,500 μH
Typ. inductance	198 μH	447 μH	609 μH	919 μH	1,361 μH
I_{sat}	1.15 A	0.8 A	0.61 A	0.55 A	0.46 A
R_{Lout}	0.57 Ω	1.25 Ω	1.90 Ω	2.70 Ω	3.80 Ω
C_{Lout}	6.34 pF	6.75 pF	7.75 pF	7.50 pF	7.75 pF
f_{sw} at $V_{in} = 110$ V	190.1 kHz	89.1 kHz	61.1 kHz	45.8 kHz	33.5 kHz
f_{sw} at $V_{in} = 325$ V	275.9 kHz	131.3 kHz	90.0 kHz	68.1 kHz	49.6 kHz

one used for the characterizations in Sects. 5.1–5.3 is a $12 \times 12 \times 10$ mm package with a volume of 1,440 mm^3. The other available packages are $10 \times 10 \times 6$ mm package with a volume of 600 mm^3 and $7.3 \times 7.3 \times 4.5$ mm package with a volume of 240 mm^3.

Table 5.3 shows a parameter overview of the inductors in a $10 \times 10 \times 6$ mm package, which is approximately 60% smaller than the previously used inductors. For the full converter, this reduces the volume by 7.2% from 11,650 mm^3 to 10,800 mm^3. Thus, 7.8% higher power density is accomplished. The achievable increase in power density is limited by the EMI filter and input buffer capacitors, which are defined by the input conditions and cannot be influenced by the subsequent converter (see Sect. 4.5).

Of the available inductors listed in Table 5.3, only the ones with an inductance of 470 μH and below show a sufficiently high saturation current to support the required peak current $I_{L,max} > 620$ mA. The 220 μH inductor leads to an on-time outside of the specified range for the silicon converter IC (see Sect. 5.3).

For the smallest inductor size of $7.3 \times 7.3 \times 4.5$ mm, even the 220 μH inductor has not a sufficiently high saturation current. However, it is able to supply the desired output current of 300 mA by driving the magnetic material of the inductor core into saturation. Thus, the inductance value reduces when the inductor current I_L is around the desired peak current. This leads to an increased operation frequency and thereby to increased switching losses in the converters.

Figure 5.21 shows the efficiency over power density achieved by different inductor package sizes for various input voltages. The first values in each plot at a power density of 21.2 W/in^3 correspond to the largest $12 \times 12 \times 10$ mm inductor. The smallest inductor size in a $7.3 \times 7.3 \times 4.5$ mm package leads to the values at a power density of 23.3 W/in^3.

The GaN-based converter achieves higher efficiency at all examined operating conditions. The saturation of the smallest inductor leads to an increased switching frequency of 290 kHz and a reduced on-time of the power transistor. This cannot be supported by the silicon converter IC anymore, likely due to a violation of the minimum on-time required for the

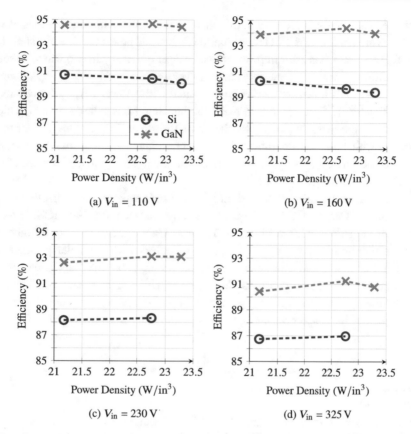

Fig. 5.21 Comparison of measured efficiency over power density between GaN and silicon-based converters with differently sized 220 μH inductors for **a** $V_{in} = 110$ V, **b** $V_{in} = 160$ V, **c** $V_{in} = 230$ V and **d** $V_{in} = 325$ V

peak current control. Hence, the silicon-based converter does not operate for input voltages above 200 V with this inductor. However, the GaN IC is able to support the higher operation frequency with an efficiency greater than 90% even at $V_{in} = 325$ V. It is remarkable that the GaN IC together with the intermediate coil size and the associated power density of 22.8 W/in^3 leads to the highest conversion efficiency over the full input voltage range.

Thanks to lower $R_{DS,on}$ with low C_{oss} as well as small I-V overlap losses and low power circuit design achieving lower supply losses (Fig. 5.19), the GaN IC is able to outperform the efficiency of this commercially available silicon converter IC at all examined operation conditions by at least 2% (Fig. 5.20). This corresponds to a power loss reduction between 25% and 42% depending on the operation point. Furthermore, due to the faster switching capability of GaN, a wider range of inductors can be utilized, which enables implementations with higher power density while maintaining high efficiency above 90% for a voltage conversion of 325 V to 47 V.

With the loss analysis performed in this chapter, the mechanisms and sources of various power losses are identified and their contribution to the total power loss is estimated. The use of GaN for the power transistor improves the converter efficiency in two ways, directly and indirectly. Due to the approximately ten times lower figure of merit $R_{DS,on} \cdot Q_{oss}$ compared with silicon, the GaN transistor can be implemented with a lower $R_{DS,on}$ directly reducing the conduction losses. Concurrently, a smaller output charge has to be dissipated, which also directly reduces the switching losses. Furthermore, the use of GaN enables a wider selection of inductors which allows to optimize the conduction and switching losses caused by the output inductor. Thereby, the use of GaN can contribute indirectly to improve the efficiency of the full converter system.

The power density is also improved directly and indirectly by the utilization of a GaN power transistor. Despite the three times lower on-resistance, the GaN IC is three times smaller than the silicon IC (see Fig. 5.2). Thereby, GaN directly leads to a higher power density. Furthermore, the use of GaN enables inductors with a smaller package size to be employed as output inductor for the buck converter. Thus, the GaN power transistor indirectly increases the power density of the converter. This indicates that the benefits of GaN enabled by the circuit and system design of this work are representative and the results and considerations of this chapter can also be transferred to other converter types and application spaces.

References

1. Power Integrations. (2016). Single-stage LED driver IC with combined PFC and constant current output for buck topology. In *LYT1402-1604 LYTSwitch-1 Family Datasheet*.
2. STMicroelectronics. (2014). Offline LED driver with primary-sensing and high power factor up to 15 W. In *HVLED815PF Datasheet*.
3. Kaufmann, M., & Wicht, B. (2020). Monolithic GaN-IC with Integrated Control Loop Achieving 95.6
4. Power Integrations. (2019). GaN-based primary-side power switches extend the power range of Innoswitch3 IC families. https://ac-dc.power.com/design-support/articles/powigan-based-primary-side-powerswitches-extend-power-range-innoswitch3-ic-families/. Retrieved from 04/06/2021.
5. Wuerth Electronics. (2019). WE-PD HV SMT Inductors PSpice simulation model. https://www.we-online.de/katalog/download/PSpice_WEPD_HV. Retrieved from 03/19/2021.
6. LeCroy. (2005). PP008 passive probe. In *Instruction Manual Revision A*. https://docs.rs-online.com/916f/0900766b80f74307.pdf. Retrieved from 03/19/2021.
7. Rohm Co., Ltd. (2017). RFN2LAM6STF super fast recovery diode. In *Datasheet Rev. B*. https://www.mouser.de/datasheet/2/348/rfn2lam6stf-e-1869648.pdf. Retrieved from 03/19/2021.
8. Cree, Inc. (2013). CSD01060 silicon carbide Schottky diode zero recovery ®rectifier. In *Datasheet Rev. P*. https://www.wolfspeed.com/media/downloads/87/CSD01060.pdf. Retrieved from 03/23/2021.

Conclusion and Outlook

6

This chapter summarizes and concludes the research presented in this book. Finally, an outlook on the future of integrated GaN circuits for power conversion is provided.

6.1 Conclusion

Power converters for input voltages above 20 V are a growing application space. It includes the 48 V input rail common in server and automotive applications and extends to converters operated directly from the rectified mains voltage in various applications including chargers and LED lighting. For all of these applications, high-power density is desired to reduce the volume and weight of the required power converters. This demands for faster operation frequencies of switching converters in order to reduce the size of bulky passive components. Especially for mass market applications such as chargers and LED lighting, a high level of integration is desired to obtain easy-to-use converters with low external component count and low PCB complexity. Concurrently, high power efficiency is required in order to minimize the self-heating of compact power converters as well as to reduce the global power consumption.

Traditional power converters using silicon transistors reach their switching frequency limit below 100 kHz. This limits the minimum values and sizes for capacitors and inductors, consequently restricting the achievable power density. Hence, wide bandgap semiconductors with superior figures-of-merit for high-voltage and high-frequency switching operations experience an increasing interest in the application in power converters. Thereby, higher power density and concurrently higher conversion efficiency can be achieved. GaN shows the best theoretical material limit (Baliga's figure-of-merit) of the established materials leading to small high-voltage power transistors with small specific capacitances. Furthermore, price parity between GaN and silicon on device and system level is expected soon. Hence, GaN is a very promising material to obtain compact power converters with high efficiency.

© The Author(s), under exclusive license to Springer Nature Switzerland AG 2022   167
M. P. Kaufmann and B. Wicht, *Monolithic Integration in E-Mode GaN Technology*,
Synthesis Lectures on Engineering, Science, and Technology,
https://doi.org/10.1007/978-3-031-15625-0_6

Nevertheless, driving of GaN transistors is challenging due to tight gate voltage requirements. Typically, they allow a maximum transient gate voltage of 10 V and show a threshold voltage as low as 1.1 V. Hence, the parasitic inductance of the gate loop has to be minimized to values below 1 nH in order to avoid harmful gate voltage overshoots and unwanted turn-on of power transistors during switching transitions with slew rates up to 200 V/ns. Thereby, simple and cost-effective unipolar gate drivers without additional protection such as clamping diodes can be used. Appropriately, today's GaN transistors are formed by a lateral structure. This allows for monolithic integration of power transistor and driver on one IC reducing the gate loop inductance to values close to zero as demonstrated in various publications. In order to further reduce the system complexity and the number of components on PCB, monolithic integration of additional functions such as control loops, sensing circuits, supply voltage generation, and protection features are required. Therefore, this work investigates the possibility and viability of monolithic integration of power converter ICs in an e-mode GaN technology.

Publications on e-mode GaN transistors and technologies are summarized, focusing on opportunities for integrating circuits in an e-mode GaN technology based on a p-GaN Schottky gate structure. In GaN, the hole mobility is typically more than 50 times smaller than the electron mobility, leading to at least 50 times weaker p-type devices. This lack of suitable complementary devices poses various challenges for circuit design where neither efficient and fast CMOS logic nor p-type current sources as load for amplifier stages can be implemented. Furthermore, cost-effective GaN grown on silicon wafers shows a comparably high crystal defect density, which is even increased by the p-GaN layer in state-of-the-art e-mode GaN processes. Thus, matching of devices is challenging. Several charge trapping effects lead to dynamic R_{DSON} behavior as well as threshold voltage instabilities indicating weaker noise performance of GaN transistors in typical analog operating conditions. However, the absence of bipolar junctions eliminates reverse recovery and associated losses contributing to high-efficient switching. Furthermore, the natural isolation of devices in GaN enables effortless integration of functionality including various high-voltage circuits on one die. One part of this work is to investigate circuit design techniques to exploit the benefits of GaN technology as well as to mitigate the technology limits on circuit and system levels.

In this work, a 650 V e-mode GaN-on-Si process is assessed for its integration capabilities based on wafer-level characterization of fundamental circuits. The established small-signal model of silicon MOSFETs is also valid for GaN transistors, even though the mechanisms of the transistor functionality are different. The typical intrinsic gain of GaN transistors used in this work is characterized to be around 350 V/V, similar to silicon transistors. However, the achievable single-stage gain is reduced to values below 15 V/V since resistors are used as a load for amplifiers in this work due to the lack of suitable complementary transistors. As confirmed by wafer-level measurements, adjacent resistors can be designed to achieve a standard deviation of the mismatch below 0.8%. A study of multiple, differently sized transistors shows that Pelgrom's matching law for silicon transistors (mismatch is proportional to $1/\sqrt{W \cdot L}$) is also applicable for GaN transistors. However, the characterized absolute threshold voltage mismatch between adjacent GaN transistors is nearly two decades higher than for silicon transistors with the same area $W \cdot L$, as expected from the larger defect density published by various research groups. Thus, the offset of a differential pair with common centroid layout and dummy transistors is characterized in this work to be in the range of ± 200 mV at a standard deviation of $\sigma = 70$ mV. Noise measurements are performed for stand-alone transistors and resistors as well as for common-source amplifiers. They reveal large low-frequency drain current noise of integrated GaN transistors utilized in this work with a spectral current noise density $S_{ID} \sim 10 \times 10^{-15}$ A^2 Hz^{-1} at 10 Hz when the DC bias is at a typical value of 1 $\mu A\mu m^{-1}$ transistor width. This leads to an output noise in the range of ± 20 mV characterized for a common-source amplifier.

The lack of suitable p-type devices significantly limits the digital performance of GaN. Using 2DEG resistors with a relatively low sheet resistance of $500\Omega/\square$ as pull-up, the area required for a low-power inverter is approximately 50 times larger than for a CMOS inverter. The average propagation delay of an inverter is characterized using a ring oscillator designed as part of this work. It is in the range of 3 ns, which is more than two decades slower than for a 130 nm silicon CMOS technology. Together with the DC current caused by the resistor pull-up, this leads to a power-delay product above 4×10^{-13} Ws for the resistor–transistor logic (RTL) inverters used in this work, which is about four decades higher than for 130 nm silicon. Nevertheless, integration of analog, digital, mixed-signal, and high-voltage circuits is possible and a monolithic GaN IC is developed to demonstrate the chances of GaN integration for power electronics.

As part of this work and related publications, the first GaN buck convert IC with analog control loop suitable for offline operation is presented. A single supply rail-to-rail gate driver using bootstrapping techniques is integrated together with a 1Ω, 650 V power transistor. It achieves symmetrical propagation delays of 80 ns characterized at room temperature for both turn-on and turn-off transitions. Due to the large temperature coefficient of -5500 ppm/K for the electron mobility in GaN, the propagation delay varies by a factor of five in the temperature range from -40 °C to $+150$ °C. A first-order loss model identifies quasi-resonant mode (QRM) and discontinuous conduction mode (DCM) as suitable operation modes to achieve high conversion efficiency for the given application space with high voltages $V_{in} = 400$ V, $V_{out} = 50$ V and output power up to 30 W. QRM is chosen as operation mode due to lower complexity in the circuit-level implementation, what is well suited for integration in today's GaN technology. Furthermore, higher power density can be achieved by utilizing smaller inductors enabled by the smaller inductor peak current in QRM. Thus, the analog control loop contains an integrated, capacitive falling edge detection at the switching node to determine the moment of zero inductor current triggering the turn-on of the power transistor. The output current is regulated using a cycle-by-cycle peak current control. Hence, a peak current comparator with input referred auto-zero loop is designed as part of this work. The auto-zero input capacitors are employed to fulfill two functions simultaneously. They are used to reduce the input referred offset to less than 5 mV, while the differential pair shows an offset of up to 200 mV. Furthermore, they provide input level shifting to handle low-voltage ground referred current sense signals with an n-type input stage. However, the auto-zeroing loop is not able to reduce the noise of the comparator, which shows a standard deviation of 18.2 mV likely caused by intrinsic transistor noise. This may be addressed in future research by noise reduction techniques such as chopping as well as technology improvements.

The GaN IC achieves self-biased offline operation without requiring an auxiliary inductor winding or a supply regulator on PCB. This is enabled by an integrated high-voltage supply regulator developed as part of this work. It generates the 6 V supply voltage for the IC directly from the high-voltage switching node and provides a charging current of up to 8.5 mA. The voltage ripple is below 45 mV if an auxiliary capacitor of 470 nF is used. The integrated regulator includes a voltage loop tracking PVT variations in order to provide the highest suitable supply voltage to drive the power transistor without harming the sensitive gate. With these circuits integrated into GaN, a buck converter is implemented to conduct system-level characterization of the proposed power converter IC.

A system-level characterization of the converter confirms operation between 85 V and 400 V input suitable for application at both 110 V and 230 V power grids. The peak efficiency achieved in this work is 95.6% and a conversion efficiency above 90% is maintained for the full nominal input voltage range of 85 to 325 V. As part of this work, a measurement series with various power inductors is performed for different input voltages. For comparison, the same measurements are also carried out for a silicon converter IC applicable for similar input and output conditions. The proposed GaN IC achieves up to 42% lower power loss than the silicon-based converter. Higher switching frequencies up to 275 kHz are reached, enabling the utilization of smaller inductors not supported by silicon-based converter ICs. Thereby, the achievable power density of the full system can be increased by 7.8% to 22.8 W/in^3 at an efficiency greater than 91% for the full nominal input voltage range 85 to 325 V of universal offline power converters. This work analyzes and discusses the different loss mechanisms of the proposed GaN and the characterized silicon converter. Hence the efficiency benefit of the implemented GaN-based converter is representative and the considerations and results of this work can be transferred to other GaN converters and applications spaces. This way, the presented GaN IC shows the viability of monolithic GaN integration as a path towards compact and efficient high-voltage power supplies.

6.2   Outlook

The opportunities for monolithic integration of GaN power converter ICs are demonstrated by the design proposed in this work, which outperforms state-of-the-art silicon-based converters with respect to conversion efficiency and power density. However, due to several process-related effects, monolithic GaN ICs do not exploit the full potential of the technology, yet. In order to compensate for the lack of suitable complementary devices in state-of-the-art e-mode GaN processes, this work proposes design techniques mainly based on bootstrapping and capacitive level shifting. While this generally enables the implementation of all required functions, the performance of those circuits is limited. When the recently published fundamental p-type devices are developed further to achieve lower specific resistance compared to n-type devices, the performance of integrated circuits in GaN can be significantly improved. Comparable to the development of silicon CMOS technology, complementary GaN circuits may be a game changer for low-power digital circuits in GaN, efficiently extending the integratable computation power. Furthermore, utilizing p-type current sources as biasing circuits with high output resistance would increase the achievable single-stage voltage gain significantly. It would also allow for a larger input common mode range of core circuits such as comparators and operational amplifiers.

One limitation this work discovered is the higher drain current noise level of GaN transistors in integrated analog circuits. Further research in the area of the process technology is required in order to identify the root causes and reduce the noise levels of GaN transistors enabling analog circuits with higher precision. The application of silicon circuit design techniques for low noise, such as chopping, should also be investigated to allow for analog control loops with higher precision.

The reference voltage for the peak current control presented in this work is provided externally. In order to achieve GaN power converter ICs with an integrated voltage reference as it is common in silicon technologies, an integrated voltage reference would be required in GaN. While few publications propose different circuits for such voltage references, the performance in terms of absolute accuracy is limited and not competitive with the established silicon bandgap reference circuit.

As a first investigation vehicle, the power converter in this work is applied to LED lighting and thus provides thus an output current control. An integrated voltage loop is desirable to extend the application space of GaN power converter ICs also to general purpose converters. Therefore, the possibility of integrated high-gain operational amplifiers has to be investigated. The chances to design suitable amplifiers would be dramatically increased when the process technology offers complementary devices as well as low-noise transistors.

Today's cost-efficient GaN-on-Si technologies do not offer substrate isolation and, hence, do not support the integration of a 400 V half-bridge on one die. Few pioneering technologies offer the possibility of local substrate biasing with GaN-on-SOI technology at the cost of an increased process complexity. These technologies are limited to 200 V transistors and are therefore not suitable for half-bridges in offline converters supplied by the 230 V power grid. Advances of this technology towards higher breakdown voltages of 600 V and more would enable single-die power converter ICs for various half-bridge-based topologies.

When the technology and circuit design challenges are solved, high-performance GaN ICs can replace silicon technology in high-voltage power converters for input voltages at least up to 400 V. Various topologies such as active-clamp flyback for sub 75 W chargers without power factor correction and totem pole PFC converters for above 75 W applications may be possible using monolithic GaN. Thereby, the efficiency and power density of power converters in a huge application space can be increased, while, concurrently, the ease-of-use of established silicon solutions as well as the low bill-of-materials for full converters is maintained. Thereby, GaN technology can leverage its full potential to make an impact on the global challenge of climate change by improving the efficiency of electric power consumption and reducing greenhouse gas emissions.

Index

© The Editor(s) (if applicable) and The Author(s), under exclusive license to Springer
Nature Switzerland AG 2022
M. P. Kaufmann and B. Wicht, *Monolithic Integration in E-Mode GaN Technology*,
Synthesis Lectures on Engineering, Science, and Technology,

Printed in the United States
by Baker & Taylor Publisher Services